The Symmetries of Things

The Symmetries of Things

John H. Conway
Heidi Burgiel
Chaim Goodman-Strauss

CRC Press
Taylor & Francis Group
Boca Raton London New York

CRC Press is an imprint of the
Taylor & Francis Group, an **informa** business

AN A K PETERS BOOK

Editorial, Sales, and Customer Service Office

A K Peters, Ltd.
888 Worcester Street, Suite 230
Wellesley, MA 02482
www.akpeters.com

Library of Congress Cataloging-in-Publication Data

Conway, John Horton.
  The symmetries of things / John H. Conway, Heidi Burgiel, Chaim Goodman-Strauss.
       p. cm.
  Includes bibliographical references and index.
  ISBN-13: 978-1-56881-220-5 (alk. paper)
  ISBN-10: 1-56881-220-5 (alk. paper)
  1. Symmetry (Mathematics) 2. Geometry. 3. Shapes. I. Burgiel, Heidi, 1968- II. Goodman-Strauss, Chaim, 1967- III. Title.

QA174.7.S96C66 2008
516'.1–dc22

                                                                          2007046446

Printed in India
12 11 10 09 08   10 9 8 7 6 5 4 3 2 1

To
Gareth, Diana,
Eli, Zoe, Kendall, and
the memory of Gay Lorraine Burgiel

Contents

Preface

This book has been germinating for a long time. John Conway has always been interested in geometrical groups, for many of which he devised particular notations when he was teaching at Cambridge University. However, after he moved to Princeton University in 1985 and Bill Thurston told him of the orbifold idea, he dropped those notations forever and devised the signature notation used in this book. He then became Thurston's most avid prophet, lecturing on the theory to scores of audiences—ranging from the Princeton Rug Society to the International Congress of Mathematicians!

One of those audiences contained the young graduate student Heidi Burgiel, who was taking notes on the talk for distribution during the conference. Heidi went on to complete a graduate program in combinatorics and discrete geometry. Years later, when John spent some time at Northwestern University, Heidi offered to "write something else up" with John, but in the end they decided to write the same theory in more detail as a proposed book. That book has been growing ever since.

All they had intended to write was the content of what is now Part I—an elementary introduction to the orbifold signature notation. But then came the idea of writing a second part that would extend the signature to color symmetry. At this point it became clear that Chaim Goodman-Strauss would make an excellent addition to the team of authors. Chaim had been preaching the gospel of the orbifold signature on his own and was known for his gorgeous illustrations.

More topics burst into bloom at various seasons. When Conway, Delgado, Huson, and Thurston used the signature to re-enumerate the three-dimensional space groups, it seemed a good idea to incorporate this also in the second part. That "second part" is now Parts II and III.

Much of the book was written in hectic three-day sessions on the few occasions when all three of us could get together—this paragraph is being finalized on the way to the Tampa airport, days before the book is sent to press. We usually managed to write several chapters in each session, often including one that only arose just then. For example, at one session, Chaim said "we could perhaps do Heesch types," and an hour later Chapter 15 was complete. Just after completing the next section of this introduction (which describes what's new to this book), the three of us celebrated at a restaurant, discussed "Archimedean tilings," and Chaim and John discovered the "Archifold notation" that characterizes such things as they walked home after the meal. The next day this too was in the book. Of course, it often took Chaim years to catch up with the illustrations.

What's New in This Book?

Many of the results and proofs in this book are new, or nearly new, in the sense that their only previous appearances have been in the scholarly papers (often involving one of us) that are cited in the appropriate chapters. These new things are

- the orbifold signature,

- the statement of our Magic Theorem,

- its use to enumerate symmetry types
  (however, we should point out that a few decades ago, MacBeath introduced his own signature that is in fact equivalent to ours— but more complicated—and used it in the same way),

- Conway's "zip proof" of the classification of surfaces,

- uniform presentations for all the groups,

- their proof,

- our analysis and notation for color symmetry,

- the p-color types for all primes p,

- the simplified enumeration of Heesch types,

- the Besche-Eick-O'Brien table of group numbers,

- the extension of all of the above geometrical theory to hyperbolic groups,

- a new proof of the abstract distinctness of infinite groups with compact orbifolds,

- the explanation of isospectral "drums" via hyperbolic groups,

- the classification of Archimedean tilings in the hyperbolic plane,

- generalized Schläfli symbols,

- Architectonic 3-tessellations,

- the new space group notations and a panoply of objects with prime space symmetries,

- names and enumeration of platycosms,

- a list of Archimedean 4-polytopes.

Even when the results are old, our exposition is new.

We are also proud of our exposition and illustrations. Chaim Goodman-Strauss assures our readers that his software and illustrations are available for sale and licensing.

We are relieved that now the book is in print, bringing the orbifold signature to the world. This would not have happened without help from many people including Robert Strauss, Troy Gilbert, Marc Culler, Tom Moore, Charlotte Henderson, Alice and Klaus Peters, Bill Thurston, Silvio Levy, Peter Doyle, Natasha Jonaska, Daniel Huson, Olaf Delgado Friedrichs, Doris Schattschneider, Marjorie Senechal, Javier Bracho, our students, and our colleagues; and the patience and sympathy of our partners Diana, Kendall, and Rachel. We thank the institutions that supported our work, including Princeton University, the University of Arkansas, the Universidad Nacional Autómata de México, Northwestern University, the University of Illinois at Chicago, Bridgewater State College, and the National Science Foundation.

Figure Acknowledgments

Page 70   Frieze patterns in downtown Chicago, photographs
             by Susan McBurney: 208 S LaSalle, Near Michigan
             Ave on Cross St, Red Roof Inn, Rookery, Tribune
             Tower.

Page 82   Map on a sphere, image and software by Ken
             Stephenson, combinatorics by Jim Cannon, Bill
             Floyd, and Walter Parry.

Page 134   M. C. Escher's *Symmetry Work 67* © 2007 The
             M. C. Escher Company-Holland. All rights reserved.
             www.mcescher.com

Page 135   M. C. Escher's *Symmetry Work 22* © 2007 The
             M. C. Escher Company-Holland. All rights reserved.
             www.mcescher.com

Page 152   M. C. Escher's *Symmetry Work 70* © 2007 The
             M. C. Escher Company-Holland. All rights reserved.
             www.mcescher.com

Page 153   M. C. Escher's *Symmetry Work 67* © 2007 The
             M. C. Escher Company-Holland. All rights reserved.
             www.mcescher.com

Page 180   John, Jane, and baby, clip art by anonymous artist
             from copyright-free CD packaged with Adobe CS 1,
             arranged by Chaim Goodman-Strauss.

Page 184   Paving stones in Zakopane, Poland, photograph by
             David Harvey.

Page 224   M. C. Escher's *Circle Limit IV* © 2007 The
             M. C. Escher Company-Holland. All rights reserved.
             www.mcescher.com

Page 318   Birhombohedrille, made of Rhombo blocks created by
             Michael S. Longuet-Higgins, photograph by Chaim
             Goodman-Strauss.

Page 348   Pencil hexastix, model and photograph by Chaim
             Goodman-Strauss, construction by John Conway and
             George Hart.

Page 366   Dragonfly, clip art by anonymous artist from
             copyright-free CD packaged with Adobe CS 1, ar-
             ranged by Chaim Goodman-Strauss.

All other illustrations by Chaim Goodman-Strauss.

Part I

Symmetries of Finite Objects and Plane Repeating Patterns

Introduction to Part I

Symmetries and symmetric patterns surround us throughout our lives. The aim of the first part of this book is to describe and enumerate all the symmetries found in repeating patterns on surfaces. To prove that our enumeration is accurate, we then explain the beautiful ideas from topology and algebra that form the basis for our conclusions.

We start with a problem—enumerating symmetric patterns. We then introduce tools for solving this problem and complete the enumeration. But then we are presented with a second problem—demonstrating that these tools work the way we claim, that there is a solid mathematical foundation beneath our results. Again, we solve this problem with some tools, then present the mathematics supporting the use of those tools. In this way, each chapter reduces the problems left by the preceding chapter to another problem whose solution is postponed to the following chapter.

This is a departure from the traditional practice of building a theory starting with basic principles and working toward the ultimate goal of proving some final theorem. We believe that our backward approach will be successful because it allows us to present one concept at a time, at the cost of always postponing the proof of just one thing to the next chapter. We hope also that the argument will be clearer when presented in a single logical thread, of the form $A \Leftarrow B \Leftarrow C \Leftarrow ... \Leftarrow Z$.

The first chapter is a gentle introduction to symmetry. Chapter 2 introduces the four fundamental features that we use to classify symmetry. In Chapter 3 we state our Magic Theorem and apply it to find the 17 possible types of repeating planar patterns, while Chapters 4 and 5 perform a similar service for spherical and frieze patterns, respectively. The Magic Theorem is deduced in Chapter 6 from Euler's Theorem, which is itself proved in Chapter 7. Finally, Chapter 8 gives our new proof of the classification of surfaces, and Chapter 9 illustrates the orbifolds that underlie our theory.

– 1 –

Symmetries

Every day we are surrounded by symmetric objects and patterns. From furniture to flooring, symmetry is the rule. In art, symmetry is pleasing to the eye, and the intricacies of extremely symmetric patterns can entrance an audience. In architecture, symmetric designs are attractive for yet another reason—repetition of a design element means re-use, which ultimately requires less planning and testing. In manufacturing, it is simpler, cheaper and more efficient to repeat a pattern at regular intervals. Even Nature has reasons to use symmetry in her work.

Recently, John H. Conway and William Thurston adapted Murray MacBeath's mathematical language for discussing symmetry. Now, the symmetries of a pattern can be defined by a single symbol that we call its signature: for example, 3*3, for the pattern on the left. With some practice, almost anyone with some knowledge of high-school geometry can read this signature and identify the symmetries it describes.

Etymology

The word *symmetry* is a combination of the words *sym* (together) and *metron* (measuring). The meaning of *bilateral* is, literally, two-sided.

Kaleidoscopes

The simplest signature is just * (star). A * denotes a *mirror* or *kaleidoscopic* symmetry, and a * alone means that there are no other symmetries to the figure. The pair of gryphons (right) has a single line of mirror symmetry running between them.

(opposite page) This pattern—which to a mathematician extends forever in every direction!—has reflections and gyrations.

Vesica piscis (fish bladder) is the traditional architectural name for patterns of this shape.

This *vesica piscis* (left) has signature *2•, pronounced "star two point symmetry" or, more formally, "period two kaleidoscopic symmetry fixing a point." We use stars for kaleidoscopes to suggest the star formed by the mirrors through a kaleidoscopic point. The *period* of a kaleidoscopic point is the number of mirror lines through it. In this case two lines of mirror symmetry—one vertical, the other horizontal—meet at the center of the flower. Finally, the point (•) indicates that all the symmetries fix a point.

You can probably guess that in a figure with signature *3•, three lines of mirror symmetry meet at its center, and similarly for signatures *4•, *5•, *6•, and so on. Mark the mirror lines and find the signatures of the tracery shown above.

WAVYTUM|MUTYVAW

BDECK   OXIH|HIXO
BDECK   OXIH|HIXO

Many letters of the Roman alphabet have mirror symmetry (or approximately so)! Symmetry will vary from typeface to typeface.

For your first quiz, identify the mirror lines and signatures of these lovely cut-paper snowflakes.

Gyrations

This *triskelion* (right) appears on the coat of arms of the Isle of Man. This figure looks the same in three orientations; the rotation through 120 degrees is a congruence that takes the figure to itself. A triskelion has period 3 gyrational point symmetry and signature 3∘.

The snakes in the middle of the above figure entwine with a period 2 gyrational point symmetry and so have signature 2∘. The gothic tracery patterns to the left and right have signatures 4∘ and 6∘, respectively.

Three roman letters have gyrational symmetry.

These hubcaps have gyrational symmetries, whose signatures you may identify for your second quiz.

Rosette Patterns

Obviously, we could keep going like this, generating pictures with period 37 kaleidoscopic point symmetry or period 42 gyrational point symmetry. But what else can we do?

For the finite *rosette patterns* like those on the last two pages, there are no other signatures. In a finite pattern, all symmetries of the pattern must fix (i.e., cannot move) the center of the pattern. Reflections across the center of the rosette and rotations about its center are the only symmetries that do this, so they're the only symmetries such a pattern can have.

By experimenting with different combinations of rotational and reflective symmetries, you can easily convince yourself that the types *•, *2•, *3•, *4•, ..., *N• and 2•, 3•, 4•, 5•, ..., N• are the only signatures possible for rosettes, to which we add 1• = • for the case of no symmetry.

Milan Cathedral window.

Frieze Patterns

After isolated pictures on a page, the easiest patterns to understand are those made by repeating pictures in a row. We see patterns like this in friezes, ribbons, animal tracks and fences.

Frieze patterns photographed in downtown Chicago.

The difference between frieze patterns and isolated figures is that, in addition to any reflective and rotational symmetries of the figures that make up the pattern, a frieze pattern has a translational symmetry that takes the figure to a neighboring figure. The first half of the book concerns itself with patterns of this sort, called *repeating patterns*.

The "dart and egg" frieze pattern is truly ancient; like all frieze patterns with this type of symmetry, it is created by reflecting a motif across a line of kaleidoscopic symmetry, then repeating the pair of images forward and backward along the kaleidoscope.

Make Your Own
Frieze Patterns

You can easily generate frieze patterns using symmetric letters! Here are some examples; can you make some others?

Repeating Patterns on the Plane and Sphere

Frieze patterns have "forward and back" translational symmetry. Plane patterns add translational symmetry in another direction. These patterns can extend to cover an entire page, or beyond. We see them every day on the floors and walls around us.

In order to study the symmetries of common objects like hairbrushes and furniture, we will also need to learn about the symmetries of patterns on spheres. Basketballs have two planes of reflective symmetry, as do tennis balls. But these balls also have a 2-fold rotational symmetry. A cube has nine planes of mirror symmetry, while some soccer balls have fifteen! In order to classify such patterns we will study repeating patterns on spheres.

Where Are We?

At the beginning of this chapter we found all the possible types of symmetry for rosettes—namely $\bullet = 1\bullet$, $*\bullet = *1\bullet$, $2\bullet$, $*2\bullet$, $3\bullet$, $*3\bullet$, $4\bullet$, $*4\bullet$, ... We've also introduced three categories of repeating pattern—repeating patterns in the Euclidean plane, frieze patterns, and patterns on the sphere. The focus of this book is to classify the different types of symmetry that objects in these categories can have. We've told you roughly what it means to say that two things have the same type of symmetry, but we'll have to postpone a precise definition of our problem until we've nearly solved it.

In fact, our book will have about as many postponements as chapters! For example, in the next chapter we'll introduce four features that in fact determine the notion of symmetry type, but will postpone the proof that they do so. These features determine the signatures that we use in Chapters 2–5 and 17 to list all possible types for each of our three categories. To do so, we employ a "Magic Theorem" whose proof is postponed to Chapter 6. In that chapter we also see that the signature really describes a topological surface called an orbifold that encapsulates all the symmetries of a pattern. The Magic Theorem is then revealed to describe a simple invariant, the Euler characteristic, of this orbifold; a detailed investigation of the Euler characteristic is in turn postponed to Chapter 7. An orbifold is a special kind of surface, and our last postponement is the fact that Euler's characteristic really does characterize the different possible topological types of surface. Our new "zip proof" of this wraps up the proof of all our results, and closes the first part of our book.

– 2 –

Planar Patterns

In this book we help you understand the symmetries of things. In this chapter we look at some repeating patterns and introduce you to the way we think about them. We describe the four fundamental features of a repeating pattern in the plane (or on any surface!) and introduce the signature we use to record these features of the pattern.

Mirror Lines

The floral pattern to the left has many symmetries. For example, the pattern is left-right symmetric: it has the vertical *mirror line* shown on the left below.

The figure in the middle shows another mirror line, which is a different kind because, unlike the first one, it runs *between*, rather than *along*, the petals. Drawing all the mirror lines we can, we get the figure on the right, which is at first sight rather confusing.

Fortunately, the small part we've highlighted in the margin contains enough information to reconstruct the whole pattern. This is because if we surround this small triangle by mirrors, as in a kaleidoscope, the reflections of the original triangle will fill in the neighboring triangular regions. The reflections of these reflections will fill in the neighbors of these neighbors, and so on, until the entire pattern is restored. With three small pieces of mirror (available at most hardware stores) and a little dexterity, you can try this yourself!

The patterns of Figures 2.1 and 2.2 are less ornate. The new patterns are somewhat simpler but have all the symmetries of the original; for our purposes all three patterns are identical.

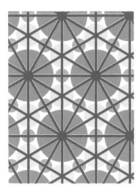

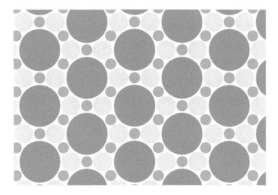

Figure 2.1. A simpler pattern.

Figure 2.2. Another simple pattern.

Repeating patterns like the ones studied in this book are made up of many symmetric copies of a motif. What we are studying here are the symmetries relating each motif to each other motif in the pattern.

Describing Kaleidoscopes

Patterns whose symmetries are defined by reflections are called *kaleidoscopic* because of their similarity to the patterns seen in kaleidoscopes. They are classified by the way their lines of mirror symmetry intersect. So, for instance, in Figure 2.3 there are three particularly interesting kinds of point, one where six mirrors meet, one where three mirrors meet, and one where two mirrors meet. We call these 6-fold, 3-fold, and 2-fold kaleidoscopic points, respectively, because the local symmetries (right) are *6•, *3•, and *2•. The whole pattern has kaleidoscopic symmetry of signature *632, where there is no final point (•) because the symmetries don't all fix a point.

Figure 2.3.   A kaleidoscope of type *632.

*632

The numbers defining the type (or signature) of a kaleidoscope can be cyclically permuted, so that *632, *326, and *263 mean the same, or also reversed, equating these with *236, *362, and *623.

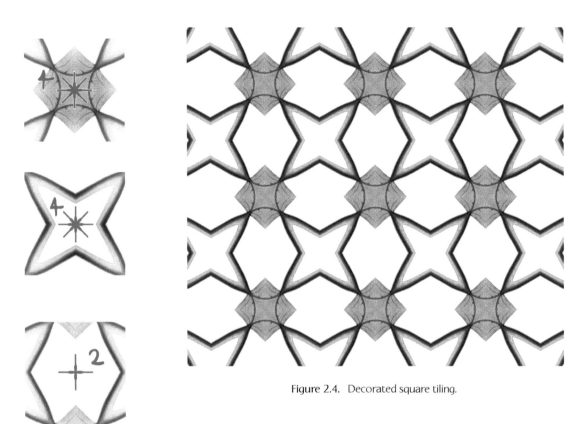

Figure 2.4. Decorated square tiling.

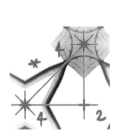

Patterns with a squarish sort of symmmetry, such as in Figure 2.4 are more common. The symmetry of this pattern is kaleidoscopic with signature ∗442. There are two 4's in the symbol because there are two different kinds of 4-fold kaleidoscopic points. The 2 in the symbol refers to the 2-fold kaleidoscopic point.

The fact that there can be several different kinds of kaleidoscopic points of the same order forces us to make it clear what *same kind* means for such points. We say, more generally, that any two features of a pattern are of the same kind only if they are related by a symmetry of the whole pattern. The points shown in the top two marginal figures are both 4-fold kaleidoscopic points but are obviously different. We will say that two points P and Q are the same if P can be moved to Q without changing the pattern's appearance in any way. (This "move" could include a reflection.)

Gyrations

The type of the kaleidoscope in Figure 2.5 is only *3 rather than
*333, because all the kaleidoscopic points in that figure are of the
same kind. However, the symmetries of this pattern are not purely
kaleidoscopic. There is a new feature—a 3-fold rotational symmetry
shown at right below.

Let's look at this more closely. The pattern would be undisturbed
if the whole plane were to be rotated through 120 degrees around
the point marked 3 in the middle of the figure. The same is also
true of the point 3 in the top figure, but we've already accounted for
this by calling it a 3-fold kaleidoscopic point—this rotation is "done
by mirrors." Since the pattern has one kind of 3-fold gyration point
and a kaleidoscope with one kind of 3-fold kaleidoscopic point, its
signature is 3*3.

Figure 2.5.   A pattern with signature 3*3.

Figure 2.6. Pattern with signature 2∗22.

2∗22

The pattern in Figure 2.6 has two kinds of 2-fold kaleidoscopic points and one kind of 2-fold gyration point. The signature of this pattern is 2∗22.

The ∗ designating the presence of a kaleidoscope separates the digit representing the gyration point from those describing the kaleidoscopic points, which are read around the kaleidoscope.

Once you are familiar with this notation, you can tell immediately that the symbol 4∗2 describes a pattern with one kind of 4-fold gyration point and one kind of 2-fold kaleidoscopic point. Figure 2.7 and the marginal figures show an example of such a pattern.

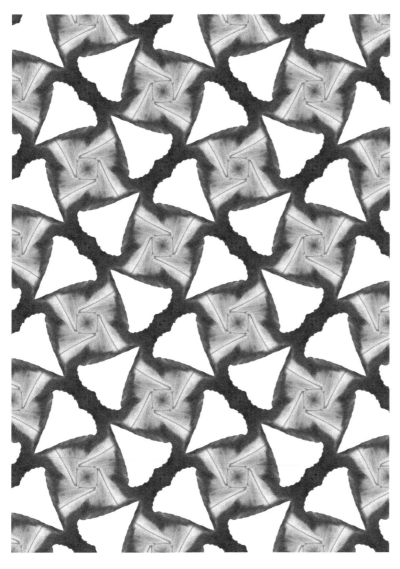

Figure 2.7. Pattern with signature 4∗2.

In Figure 2.8 we see a pattern that has only gyration points and no kaleidoscopes. Since there are three kinds of 3-fold gyration point, the symmetry is of type 333.

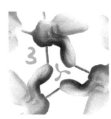

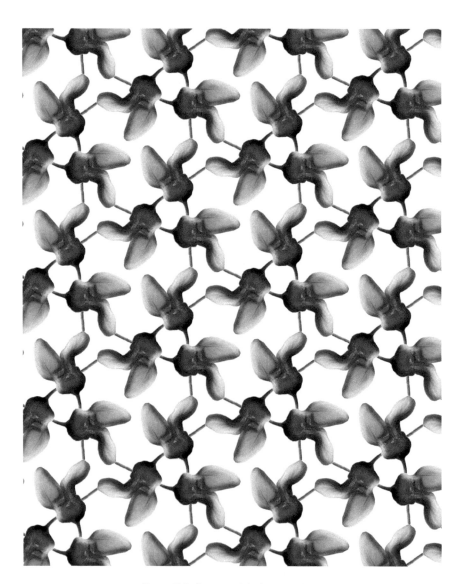

333

Figure 2.8. Pattern with signature 333.

More Mirrors and Miracles

So far we have discussed two features of patterns in the plane: kaleidoscopes and gyration points. It is natural to ask in what ways these can occur in planar patterns. For instance, can a pattern have more than one kaleidoscope?

Figure 2.9.   More than one kind of mirror signature ∗∗.

All the kaleidoscopes that we've seen so far have been defined by polygons enclosing part of our pattern, but that's not the only type there is. A single mirror line that has no other mirror lines crossing it is a kaleidoscope with signature ∗. Figure 2.9 shows a pattern with two of this kind of kaleidoscope in it, and its signature is ∗∗. (You should check that these two mirror lines really are different!)

We're also seeing something else for the first time here. The smallest subregion marked off by mirror lines in Figure 2.9 is infinite! There are several new features to be found in patterns like this, which will be presented in this section and the next.

Figure 2.10. A pattern with a mirror and a miracle: signature ∗×.

At first, Figure 2.10 looks very much like Figure 2.9. None of its mirror lines intersect, and the smallest subregion bounded by mirror lines is again infinite. But in this figure there is only one kind of mirror line!

And, there's a miracle here! There is a path from a left-handed spiral to a right-handed spiral that does not go through a mirror line. We will record the presence of such a path by a red cross (×) in the signature. We call this a "mirrorless crossing," or, for short, a *miracle*, and indicate it in figures by a red dotted line and cross.

Figure 2.10 has both mirrors and miracles, but only one kind of each, so its signature is ∗×.

We can have two miracles, just as we can have two different kinds
of mirror. This happens in Figure 2.11, which has signature ××.
(There are more than two paths from left-handed to right-handed
spirals, but all of them can be made up of combinations of identical
copies of the ones we've marked in the margin.)

Figure 2.11. More than one kind of miracle: signature ××.

Wanderings and Wonder-Rings

Just as a miracle is a repetition-with-reflection of a fundamental region that's not "explained by" mirrors, it's possible to have a fundamental region repeated without reflection in a way that's not explained by gyrations, mirrors, or miracles. In fact, such repetitions always come in pairs. We call such a pair of paths a "wonderful wandering" and denote it by a blue "wonder-ring," ○. As in the figure in the margin, we draw such a pair of paths with blue dotted lines and with a blue ring nearby. The signature for Figure 2.12 is just ○.

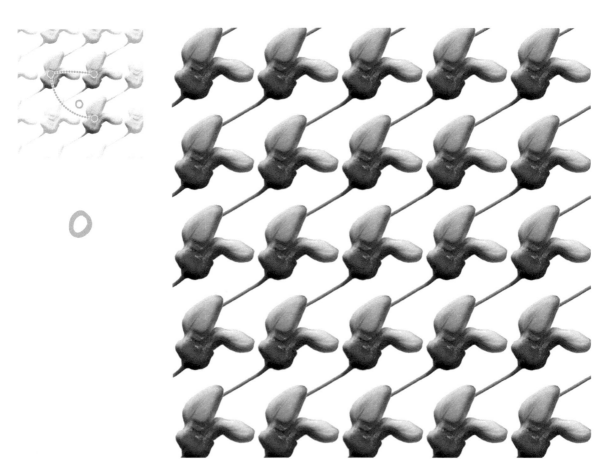

Figure 2.12. A wonderful wonder-ring: signature ○.

The Four Fundamental Features!

It is a remarkable fact that wonders, gyrations, kaleidoscopes, and mirrors suffice to describe all the symmetries of any pattern whatsoever, as we shall show in Chapter 3. We therefore call them the *four fundamental features*. You get the signature of a pattern just by writing down whichever of these features it has. Up to this point, we've used blue for wonders and gyrations, since these preserve the true orientation of a fundamental region, and red for kaleidoscopes and miracles, since these reflect. However, you can write these in black ink if you always write them in the same order, since then you'll be able to work out which colors they should be.

Table 2.1 lists the four fundamental features in the appropriate order and the codes we use to represent them in the signature.

wonders	gyrations	kaleidoscopes	miracles
○...○	AB...C	*ab...c*de...f...	×...×

Table 2.1. Features of a pattern.

Gyrations: What's in a Name?

We choose the term *gyration* to suggest motion about a point. The rotationally symmetric patterns created by crossed mirror lines are the same in the clockwise direction as they are in the counterclockwise direction. In a pattern with gyrational symmetry, there is a clear distinction between the clockwise and counterclockwise directions at the gyration point.

Where Are We?

In this chapter, we have described the four features of repeating plane patterns and introduced the *signature* that describes which of them appear in a given pattern. In the next chapter, we learn how these signatures can be used to determine what combinations of features are possible for plane patterns.

– 3 –

The Magic Theorem

In the last chapter we introduced the four fundamental features that completely describe the types of symmetry for repeating patterns. From now on we shall often specify the symmetries of a pattern just by giving its signature (which lists its features). We haven't yet said *why* just these particular features are so fundamental—and we won't, until Chapter 8—nor have we found just which signatures arise.

In this chapter we'll introduce you to the "Magic Theorem" [4], use it to show that just 17 signatures are possible for plane repeating patterns, and then deduce that such patterns come in just 17 types. The proof of the Magic Theorem itself is something else you'll have to wait for!

Everything Has Its Cost!

It turns out to be a good idea to associate a cost to every symbol in the signature, as shown in Table 3.1.

Symbol	Cost ($)	Symbol	Cost ($)
○	2	$*$ or $\times$	1
2	$\frac{1}{2}$	2	$\frac{1}{4}$
3	$\frac{2}{3}$	3	$\frac{1}{3}$
4	$\frac{3}{4}$	4	$\frac{3}{8}$
5	$\frac{4}{5}$	5	$\frac{2}{5}$
6	$\frac{5}{6}$	6	$\frac{5}{12}$
⋮	⋮	⋮	⋮
N	$\frac{N-1}{N}$	N	$\frac{N-1}{2N}$
✕	1	∞	$\frac{1}{2}$

Table 3.1. Costs of symbols in signatures.

(opposite page) The magic theorem not only classifies signatures, but helps us calculate the signature of a pattern. The signature $22\times$ of this pattern, like that of all planar patterns, costs exactly $\$2$.

Why is this? Because, as we shall see in the next few chapters, there are Magic Theorems that describe the possible signatures in terms of their costs. Here is the one we'll use in this chapter:

Theorem 3.1 (The Magic Theorem for plane repeating patterns) *The signatures of plane repeating patterns are precisely those with total cost* $^\$2$.

*632 costs $^\$2$

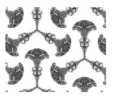

3*3 costs $^\$2$

For example, the first pattern we analyzed (Figure 2.3) had signature *632, which has cost

$$^\$1 + \frac{5}{12} + \frac{1}{3} + \frac{1}{4} = {}^\$2.$$

(Normally, we only put the dollar sign on the first of several terms to be summed.) Figure 2.5 has signature 3*3, which costs

$$^\$\frac{2}{3} + 1 + \frac{1}{3} = {}^\$2.$$

The pattern in Figure 2.6 has a kaleidoscope with two different 2-fold kaleidoscopic points and a 2-fold gyration point. Its signature is 2*22, with cost

$$^\$\frac{1}{2} + 1 + \frac{1}{4} + \frac{1}{4} = {}^\$2.$$

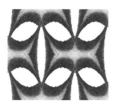

2*22 costs $^\$2$

Finally, the signature of Figure 2.9 is **, with cost

$$1 + 1 = {}^\$2.$$

This is the same as the cost for the pattern of type *× in Figure 2.10.

The proof of the Magic Theorem is quite easy, but we'll postpone it until later in our book. In this chapter we just use the theorem to help find the possible signatures for repeating patterns.

*× costs $^\$2$

Finding the Signature of a Pattern

We can now exactly identify the signature of any repeating pattern on the plane by the following steps. As we proceed, we write down the symbols in the signature, starting from the middle and working outward. If we list larger numbers before smaller ones (using *632 rather than ∗236), we can tell at a glance which patterns have the same type.

1. Mark any kaleidoscopes in red. If there are mirror lines, restrict attention to one of the regions into which they cut the plane. Put a red ∗ near any one kaleidoscope; then, find just one corner of each type (as in Chapter 2), and write the numbers of mirrors through each of these corners, also in red.

2. Look for gyration points. In blue, mark just one gyration point of each type with a spot and its order.

3. Are there miracles? Can you walk from some point to a copy of itself without ever touching a mirror line? If so, a miracle has occurred. Mark just one such path with a broken red line and a red cross nearby.

4. Is there a wonder? If you've found none of the above, then there is: mark it with a blue wonder-ring.

If you encounter a tricky pattern, there are some things you should do to make your work easier. If two features are the same you *must* only mark one of them; sometimes it helps to label gyration points before labeling kaleidoscopes. Be sure there aren't any mirror lines inside the region bounded by a kaleidoscope, and don't forget that gyration points *never* lie on mirror lines!

The rules above work for any repeating pattern; here are some more hints that work just for patterns in the plane. There is one type of plane pattern with two kaleidoscopes and one with two miracles; if you're working with one of these, you should be able to see differences between these features by looking carefully at your pattern. You know that the total cost is $2; you can use this in several ways. You can stop when it reaches $2 (for instance, if you find a wonder), or if you have not yet reached $2, you will know that there must be more features to find.

The fact that the signature of a plane pattern always costs $^\$2$ can help us check that the signature we have found for a pattern is correct: it can also help to complete it! For example, all we can see at first is that there are two kinds of 2-fold gyration points in Figure 3.1. But, 22 would only cost $^\$\frac{1}{2} + \frac{1}{2} = {}^\$1$, so there should be an extra dollar's worth to be discovered. Indeed there is! Figure 3.1 is the same as its mirror image although it has no mirror line, so there must be a miracle instead! We look at this more closely in the figure in the margin: there's a symmetry that takes a leaf to a backwards copy of itself, and the path joining these is the required mirrorless crossing, giving us the signature $22\times$.

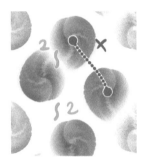

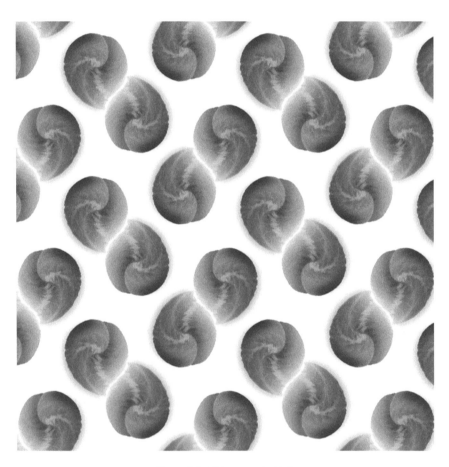

Figure 3.1. What type is this?

What is the signature of the pattern in Figure 3.2? Here there are also two kinds of 2-fold gyration points, which do not by themselves cost $^$2. The pattern is again the same as its mirror image, but a mirror, not a miracle, explains this, and the type is 22∗. (See the marginal figure.)

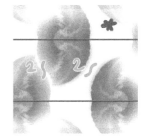

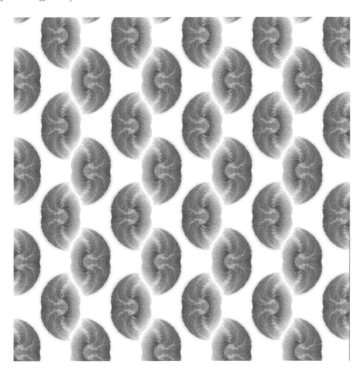

Figure 3.2.   What signature does this have?

Just 17 Symmetry Types

Why are there just 17 types of symmetry for plane patterns? We'll deduce this using only the Magic Theorem and some simple arithmetic. The calculations in the next few sections are very similar to those that answer the question, "How many different ways can I make change for a dollar if I use only quarters and dimes?" If the results at first seem mystical, try working through a few examples for yourself.

The Five "True Blue" Types

If all symmetries of a pattern are obtainable by true motions, as in the patterns on these two pages, the signature will be entirely blue. If a blue string of digits $AB...C$ is to cost $^\$2$, there must be more than two of them, since each costs less than $^\$1$. If there are exactly three, the values in Table 3.1 show that the signature can only be one of 632, 442, or 333. If there are more, it can only be 2222, since each digit costs at least $^\$\frac{1}{2}$. Finally if there's a wonder-ring, the signature must be $\circ$, since the ring already costs us $^\$2$.

The following figures illustrate the five true blue types: 632, 442, 333, 2222, and $\circ$.

Exercice

Check the types on these two pages.

We get type 333 if all characters have the mean cost of $^\$\frac{2}{3}$. Otherwise, one character must be 2.

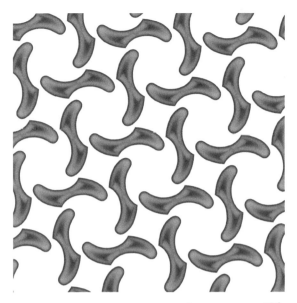

If the remaining two characters have *their* mean cost of $\$\frac{3}{4}$, we get 442.

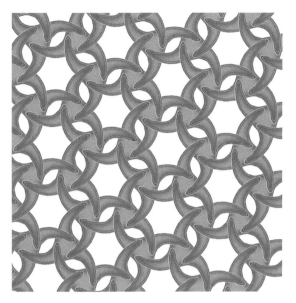

If not, a second character must be 3, and 632 is forced, since $\$\frac{5}{6} + \frac{2}{3} + \frac{1}{2} = \$2$.

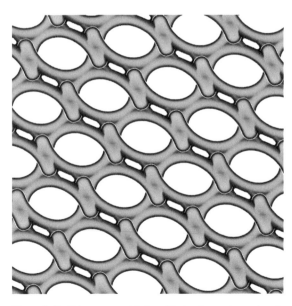

The only Euclidean type with four kinds of gyration points is 2222, since $\$\frac{1}{2} + \frac{1}{2} + \frac{1}{2} + \frac{1}{2}$ is already $\$2$.

If there's a wonder ring ○ (costing $\$2$), there can't be anything else.

The Five "Reflecting Red" Types

Now consider the signatures that are entirely red and have no crosses. They correspond to the previous cases because $*AB\ldots C$ costs $^\$2$ if and only if if $AB\ldots N$ does:

$$^\$1 + \frac{A-1}{2A} + \ldots + \frac{N-1}{2N} = {}^\$2 \iff {}^\$\frac{A-1}{A} + \ldots + \frac{N-1}{N} = {}^\$2,$$

Exercise

Check the types on these two pages.

while there can only be one such signature $(**)$ with more than one star. This yields the five reflecting red types.

The all-red signatures, ∗333, ∗442, ∗632, ∗2222, and ∗∗, corre-spond exactly to the all-blue signatures 333, 442, 632, 2222, and ○, since each red digit costs half as much as the corresponding blue digit and a kaleidoscope (∗) costs half of $2.

The Seven "Hybrid" Types

The remaining signatures either mix blue and red or involve $\times$ symbols. To help us enumerate these "hybrid" types, we note that the "demotions"

$$\text{replace } n* \text{ by } *nn$$
$$\text{replace } \times \text{ by } *$$

don't change the cost and must eventually lead to one of the five previous cases. So, we can recover all these mixed signatures by making the inverse "promotions"

$$\text{replace } *nn \text{ by } n*$$
$$\text{replace a final } * \text{ by } \times$$

in all possible ways:

$$
\begin{array}{ccccc}
*632 & *442 & *333 & *2222 & ** \\
 & \downarrow & \downarrow & \downarrow & \downarrow \\
 & 4*2 & 3*3 & 2*22 & *\times \\
 & & & \downarrow & \downarrow \\
 & & & 22* & \times\times \\
 & & & \downarrow & \\
 & & & 22\times &
\end{array}
$$

Exercise

Check the types on these pages.

The following seven figures represent the mixed types $3*3$, $4*2$, $2*22$, $22\times$, $22*$, $*\times$, and $\times\times$.

We conclude that there are just 17 possibilities for the signature, and so just 17 symmetry types for repeating patterns on the plane. (See Table 3.2.)

∗632	∗442	∗333	∗2222	∗∗
			2∗22	
				∗×
	4∗2	3∗3	22∗	
				××
			22×	
632	442	333	2222	○

Table 3.2. The 17 symmetry types of plane patterns

So indeed the Magic Theorem does imply that there are at most 17 symmetry types for a plane repeating pattern. These are traditionally called the 17 plane crystallographic groups.[1]

How the Signature Determines the Symmetry Type

We have ignored some details. To what extent can we recover the symmetry of a pattern from its signature? This is a real problem, as we shall see in the spherical case, but the answers in the plane case are easy. In the end, they depend only on the existence of rectangles and triangles with given angles, provided that those angles have the correct sum of π.

For instance, a pattern with signature ∗632 must be generated by reflections in the sides of a triangle with angles $\frac{\pi}{6}$, $\frac{\pi}{3}$, and $\frac{\pi}{2}$. All triangles that satisfy this condition will be the same up to size, so up to similarity there's just one possibility for the symmetries of a pattern with signature ∗632.

For 4∗2, four copies of a fundamental region combine to form a square. Then reflections in the sides of that square generate the rest of the pattern, as shown to the left, so there's really only one set of symmetries corresponding to 4∗2 as well. Case-by-case arguments like these work for all 17 types; you can confirm for yourself that the argument given for 4∗2 is easily adapted to the types 3∗3 and 2∗22.

---

[1]The nonreflecting elements of any of these groups form its rotation subgroup, at the bottom of the column.

In the same vein, the symmetries of a pattern with signature *2222 are generated by the reflections in the sides of a quadrilateral whose four angles are $\frac{\pi}{2}$—that is to say, a rectangle. Here the set of symmetries is no longer unique up to scale; any one version can be continuously reshaped into any other by gradually varying this rectangle.

The result is that one set of symmetries can be continuously transformed into the other while consistently maintaining its type. In technical language this kind of deformation is called an *isotopy*. So, we'll say that the symmetries of any one pattern with a given signature can be *isotopically reshaped* to become those of any other pattern with the same signature.

Interlude: About Kaleidoscopes

Kaleidoscopes—the physical kind found in toy stores—were invented by Sir David Brewster in 1816. In a real kaleidoscope, with a properly repeating, planar pattern seen at the end, the mirrors can only be arranged as shown on the right.

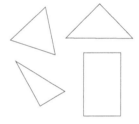

That is, the symmetry signature is just that of one of the reflecting red types *333, *442, *632, or *2222. Obtain some mirrors and make a kaleidoscope yourself!

Where Are We?

What we've shown (using the Magic Theorem, of course!) is that up to isotopic reshaping there are just 17 plane crystallographic groups. As we said, you'll have to wait to see why the Magic Theorem is true.

The next two chapters will discuss the versions of it that apply to patterns on the sphere and to planar frieze patterns.

Exercises

We've told you how to find the signature of a pattern, but most people need some practice to get it right. Follow the steps on page 31 to identify the types of the patterns on pages 42-49.

1. Repeating patterns on brick walls.

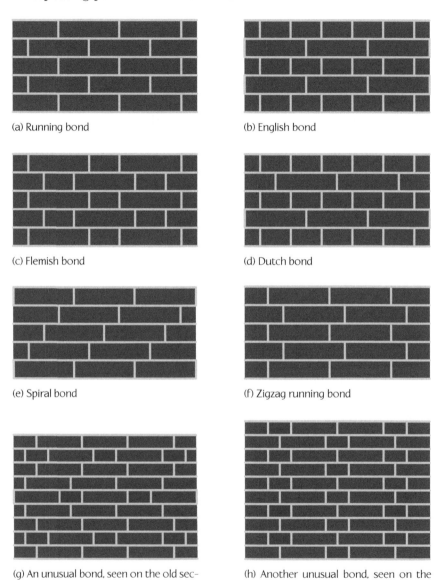

(a) Running bond

(b) English bond

(c) Flemish bond

(d) Dutch bond

(e) Spiral bond

(f) Zigzag running bond

(g) An unusual bond, seen on the old section of Princeton's Frist Student Center

(h) Another unusual bond, seen on the new section of the Frist Student Center

Check your answers.

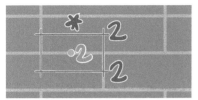

(a) Running bond has type 2∗22

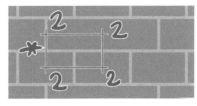

(b) English bond has type ∗2222

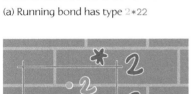

(c) Flemish bond has type 2∗22

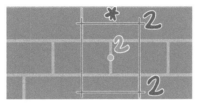

(d) Dutch bond also has type 2∗22

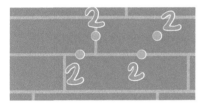

(e) Spiral bond has type 2222

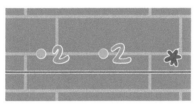

(f) Zigzag running bond has type 22∗

(g) Old Frist bond has type 22∗

(h) New Frist bond also has type 22∗

2. The placement of the dots changes the symmetry types of these patterns. Identify them.

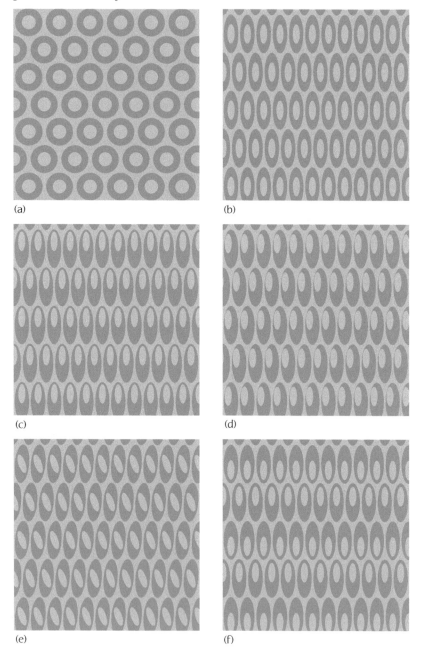

(a)

(b)

(c)

(d)

(e)

(f)

Check your answers.

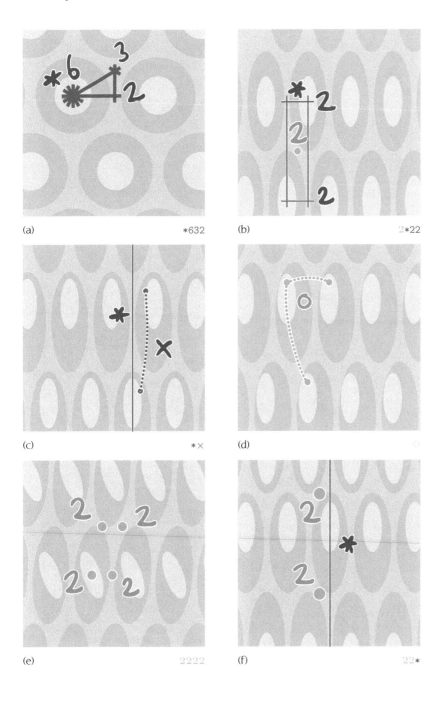

(a)   *632

(b)   2*22

(c)   *×

(d)   ○

(e)   2222

(f)   22*

3. Find the signatures of these patterns.

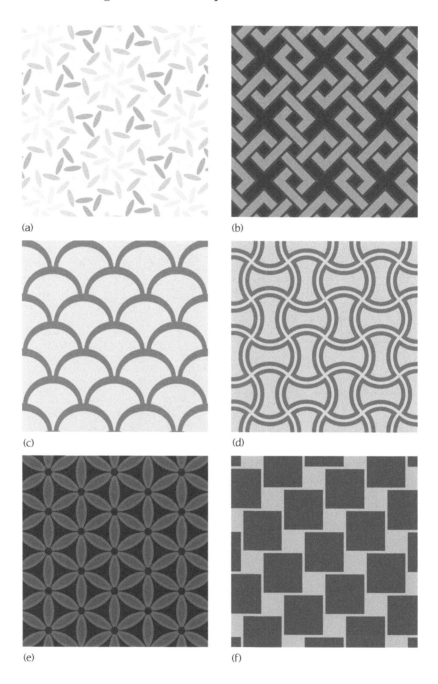

(a)

(b)

(c)

(d)

(e)

(f)

Check your answers.

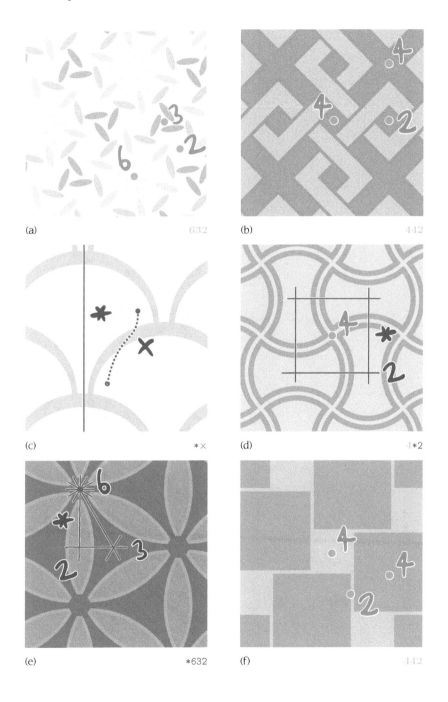

(a)   632

(b)   442

(c)   *×

(d)   4*2

(e)   *632

(f)   442

4. Even more!

(a) (b)

(c) (d)

(e) (f)

Check your answers.

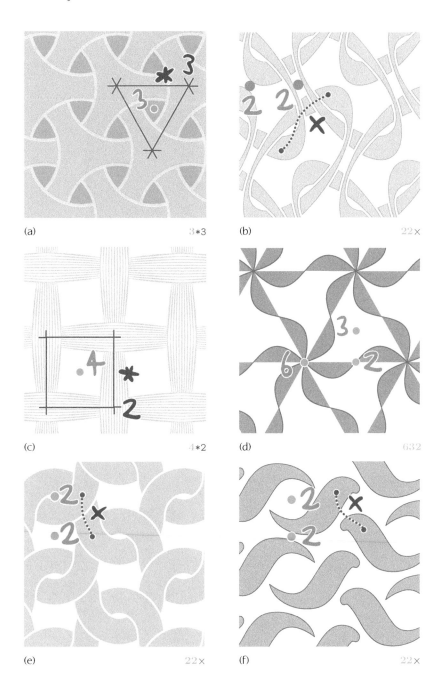

(a)   3∗3   (b)   22×

(c)   4∗2   (d)   632

(e)   22×   (f)   22×

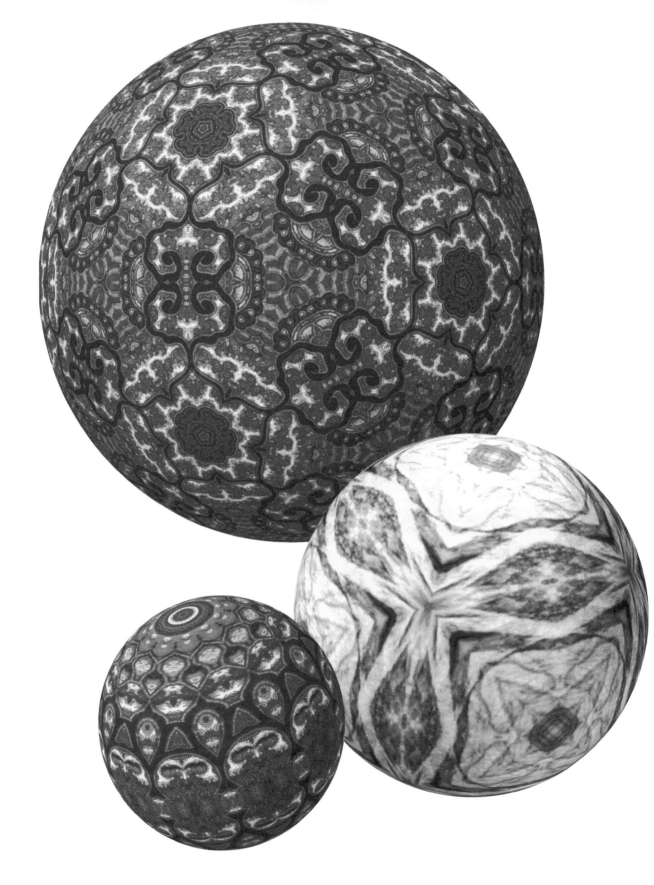

– 4 –

The Spherical Patterns

So far, we have discussed only symmetric patterns on planar surfaces. However, most of the symmetric things we encounter in our everyday lives aren't planar surfaces. Chairs, desks, boxes, and even people (roughly) are symmetric, but non-planar.

To find the features describing the symmetries of an object like a chair or table we imagine it as resting inside the "celestial sphere".

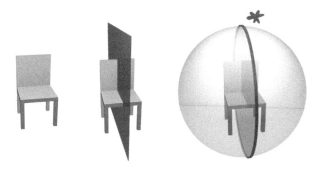

(opposite page) Three spherical patterns, with signatures *532, *2 2 11, and *432.

For the chair there is a single plane of reflection that intersects the sphere in a single mirror line—in other words, it has *bilateral symmetry*. The signature for the bilateral type of symmetry is ∗, because we see one mirror line on the surface of the sphere and it meets no other mirror lines.

We see from Table 3.1 that this only costs $^\$$1, so it is cheaper than the plane crystallographic groups, which all cost $^\$$2.

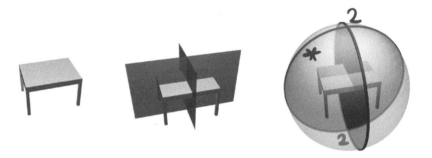

More complicated objects can have kaleidoscopic points, gyration points and miracles. For the rectangular table above, the mirror lines are two great circles that meet at right angles. On the sphere they have two intersection points, both of angle $\frac{\pi}{2}$, so the symmetries of this table have type ∗22. They cost

$$^\$1 + \frac{1}{4} + \frac{1}{4} = {^\$}\frac{3}{2},$$

again less than $^\$$2.

It turns out that an important quantity is the change we get from $^\$$2, for which we will use the abbreviation ch. Thus,

$$ch(Q) = {^\$}2 - cost(Q).$$

In particular,

$$ch(\ast) = {^\$}2 - cost(\ast) = {^\$}2 - 1 = {^\$}1,$$

$$ch(\ast 22) = {^\$}2 - cost(\ast 22) = {^\$}2 - \frac{3}{2} = {^\$}\frac{1}{2}.$$

The signatures of the Euclidean plane patterns all cost exactly $^\$2$, so if you purchased any one of them with a $^\$2$ bill, you would get no change at all. But for spherical patterns, which have only finitely many symmetries, the rule is different: the change you get is precisely $^\$2$ divided by the number of symmetries.

Theorem 4.1 (The Magic Theorem for spherical patterns) *The signature of a spherical pattern costs exactly* $^\$2 - \frac{2}{g}$, *where g is the total number of symmetries.*

In particular, the change is always positive, so the cost is always less than $^\$2$. We'll prove this in Chapter 6. In this chapter, we'll use it to derive the list of possible types of spherical pattern.

Our Euclidean Magic Theorem is really just a particular case of this, because there $g = \infty$ and so the change is $ch = {}^\$\frac{2}{\infty}$, or 0. Thus, we don't really have two magic theorems but only one.

The 14 Varieties of Spherical Pattern

Here the conclusion from the Magic Theorem is only that the spherical types are among

*532	*432	*332	*22N	*MN
				N*
		3*2	2*N	N×
532	432	332	22N	MN

but it turns out that there is a proviso: the types *MN and MN only happen when $M = N$. Here M and N represent arbitrary positive integers. We allow these integers to be 1, with the convention that digits 1 can be omitted. This makes sense—a gyration point of order 1 or a kaleidoscopic point with exactly 1 mirror passing through it is uninteresting to us, so we let 1* = *11 = *.

As in Chapter 2, we proceed by first counting the all-blue spherical signatures, then the red ones, and finally those that involve both colors.

The Five "True Blue" Types

Since the total cost of the signature must be less than $^\$2$, we cannot afford a wonder ring ($\circ$) or to have more than three digits (distinct from 1). The most general signature with fewer than three digits may be written MN by inserting 1's if necessary. Every such signature does cost less than $^\$2$, but according to the proviso it only corresponds to a symmetry type if M = N.

Finally, if there are exactly three digits, then one must be a 2, because $^\$\frac{2}{3} + \frac{2}{3} + \frac{2}{3} = {}^\$2$. Then the symbol is 22N if there are two or more 2's, and just 332, 432 or 532 if there is only one.

First note that if the signature contains two 2's, it must be 22N for some N.

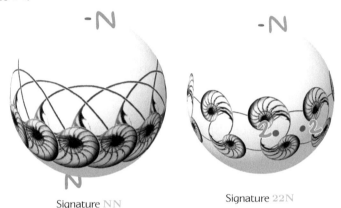

Signature NN

Signature 22N

If there is just one 2, then some other digit must be 3 since $^\$\frac{1}{2} + \frac{3}{4} + \frac{3}{4} = {}^\$2$; then, the remaining digit must be 3, 4, or 5 since $^\$\frac{1}{2} + \frac{2}{3} + \frac{5}{6} = {}^\$2$.

Signature 332

Signature 432

Signature 532

The Five "Reflecting Red" Types

The all-red signatures for sphere patterns must have the form $*\text{AB}\ldots\text{N}$ since we can no longer afford two $*$'s. The ones for which ch is positive are in perfect correspondence with the true blue types, since $ch(*\text{AB}\ldots\text{N})$ is exactly half of $ch(\text{AB}\ldots\text{N})$, as we see from the following:

$$ch(*\text{AB}\ldots\text{N}) = {}^{\$}2 - 1 - \left(\frac{A-1}{2A} + \cdots + \frac{N-1}{2N} \right),$$

$$ch(\text{AB}\ldots\text{N}) = {}^{\$}2 - \left(\frac{A-1}{A} + \cdots + \frac{N-1}{N} \right).$$

But remember the proviso: $*\text{MN}$ exists only if $M = N$.

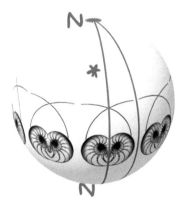

Signature $*\text{NN}$

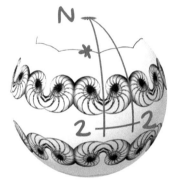

Signature $*22\text{N}$

Signature $*432$

Signature $*532$

Signature $*332$

The Four Hybrid Types

As in the plane case, these must all be obtainable by promotion from the red reflective cases. Here are all the possibilities:

$$*532 \quad *432 \quad *332 \quad *22N \quad *NN$$
$$\qquad\qquad\qquad \downarrow \qquad \downarrow \qquad \downarrow$$
$$\qquad\qquad\qquad 3*2 \qquad 2*N \qquad N*$$
$$\qquad\qquad\qquad\qquad\qquad\qquad \downarrow$$
$$\qquad\qquad\qquad\qquad\qquad\qquad N\times$$

Signature 3*2

Signature N*

Signature 2*N

Signature N×

The Existence Problem: Proving the Proviso

All 17 possibilities that we enumerated for plane patterns actually arose. In the spherical case, the corresponding statement is not quite true; the types MN and *MN only exist if $M = N$. The other cases cause no problem.

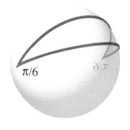

A triangle on the sphere with angles $\frac{\pi}{5}$, $\frac{\pi}{3}$, and $\frac{\pi}{2}$.

For example, *442 was generated by reflections in a triangle of angles $\frac{\pi}{4}$, $\frac{\pi}{4}$, $\frac{\pi}{2}$, and a plane pattern with this type of symmetry exists because such a triangle exists in the Euclidean plane. Similarly, *532 is generated by reflections in a triangle of angles $\frac{\pi}{5}$, $\frac{\pi}{3}$, $\frac{\pi}{2}$, and a spherical pattern with this symmetry exists because there is a spherical triangle with these angles.[1]

Now for the proviso! The type *MN, when it exists, is generated by the reflections in the sides of a two-sided polygon with angles $\frac{\pi}{M}$ and $\frac{\pi}{N}$. This *does* exist when $M = N$; it's the lune bounded by two great semicircles at angle $\frac{\pi}{N}$ (at right), but does not when $M \neq N$. (For the same reason *M, which equals *M1, fails to exist for $M > 1$.)

A two-sided spherical polygon with two angles of $\frac{\pi}{6}$ radians.

A hypothetical pattern of type MN with $M \neq N$ would contain just two types of gyration point. But then, by superposing it with its image under a reflection fixing a gyration point of each type, we should obtain one of type *MN, which is impossible. Therefore, MN also fails to exist if $M \neq N$, and M fails to exist if $M \neq 1$.

Group Theory and All the Spherical Symmetry Types

Group theory is not discussed in detail here, at least not until much later in this book; there are many texts available that teach group theory better than we are able to in the space available here. In brief, the symmetries of a pattern form a group; suppose A and B are symmetries of a pattern, described by some motion of the pattern that takes a fundamental region to a copy of itself. Then, the "product" symmetry AB is what you get when the motion associated with A is followed by the motion associated with B. You may wish

---

[1] In the plane, the sum of the angles of a triangle is always π radians $= 180°$. On a sphere, the sum can be larger. For example, a triangle on the globe with one vertex on the North Pole and two vertices on the equator has two angles of $\frac{\pi}{2}$ radians at the equator. The sum of the angles of that triangle will be π plus the angle at the pole.

Order	NN	N×	N*	2*N	22N	*NN	*22N			Number of Groups
1	11									1
2	22	×	*		22	*				3(1)
3	33									1
4	44	2×	2*	2*	222	*22	*22			5(2)
5	55									1
6	66	3×	3*		223	*33				5(2)
7	77									1
8	88	4×	4*	2*2	224	*44	*222			7(4)
9	99									1
10	10 10	5×	5*		225	*55				5(2)
11	11 11									1
12	12 12	6×	6*	2*3	226	*66	*223	332		8(4)
13	13 13									1
14	14 14	7×	7*		227	*77				5(2)
15	15 15									1
16	16 16	8×	8*	2*4	228	*88	*224			7(4)
17	17 17									1
18	18 18	9×	9*		229	*99				5(2)
19	19 19									1
20	20 20	10×	10*	2*5	22 10	*10 10	*225			7(3)
21	21 21									1
22	22 22	11×	11*		22 11	*11 11				5(2)
23	23 23									1
24	24 24	12×	12*	2*6	22 12	*12 12	*226	*332 432	3*2	10(6)
27	27 27									1
30	30 30	15×	15*		22 15	*15 15				5(2)
33	33 33									1
36	36 36	18×	18*	2*9	22 18	*18 18	*229			7(3)
39	39 39									1
42	42 42	21×	21*		22 21	*21 21				5(2)
45	45 45									1
48	48 48	24×	24*	2*12	22 24	*24 24	*22 12	*432		8(5)
51	51 51									1
54	54 54	27×	27*		22 27	*27 27				5(2)
57	57 57									1
60	60 60	30×	30*	2*15	22 30	*30 30	*22 15		532	8(4)
63	63 63									1
66	66 66	33×	33*		22 33	*33 33				5(2)
⋮	⋮	⋮	⋮	⋮	⋮	⋮	⋮			⋮
120	120 120	60×	60*	2*30	22 60	*60 60	*22 30	*532		8(5)
	C	2 × *C*		*D*		2 × *D*		*P*	2 × *P*	

Table 4.1. Types of spherical repeating patterns and their equivalences as groups.

		for $n = 3, 4, 5$
C	cyclic	$\langle a \mid a^n = 1 \rangle$
D	dihedral	$\langle a, b \mid a^n = b^2 = (ab)^2 = 1 \rangle$
P	polyhedral	$\langle a, b, c \mid a^n = b^3 = c^2 = abc = 1 \rangle$
2 × G	direct product of G with a group of order 2.	

Table 4.2. Legend for Table 4.1.

to confirm for yourself that in the examples we have seen so far this composition is associative (i.e., that $(AB)C = A(BC)$ for all choices of A, B, and C) and that in the relation $AB = C$, any two of A, B, and C uniquely determine the third.

The size or *order* of a group is the number of elements in it—here it is the number of symmetries. Two different geometrical groups have the same abstract structure if their elements multiply in the same way.

All the Spherical Types

Table 4.1 lists the types of spherical repeating patterns and their equivalences as groups. The table is complete for patterns with up to 24 symmetries, after which we restrict to multiples of 3. The last column gives the number of distinct geometrical groups, followed in parentheses by the number of different abstract structures (as separated by the lines). Each row of Table 4.1 lists all the groups of a given size; groups with different abstract structures are separated by the curved and straight lines. Codes for these structures are given below the table and are explained in Table 4.2: the polyhedral groups are isomorphic to the alternating and symmetric groups A_4, S_4, or A_5 according as n is 3, 4, or 5.

Where Are We?

We have shown in this chapter that the spherical form of the Magic Theorem implies that the spherical symmetry types fall into seven infinite families plus seven individual types. The groups of symmetries of these patterns are listed by increasing number of symmetries in Table 4.1. As before, the proof of the Magic Theorem that leads to this conclusion is postponed to Chapter 6.

Examples

Polyhedra

This marked cube has type 332; an un-marked cube has type *432.

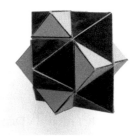

The cube and octahedron, and their "marriage," shown here, all have type *432.

This marked octahedron has type 3*2.

The snub dodecahedron has type 532.

The regular tetrahedron has type *332, and a regular icosahedron has type *532. However, this model has type 332.

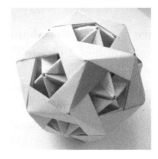

If we ignore the colors, this origami model has type 532; the symmetry type of each (one-colored) band is 225. There is no symmetry that fixes all the colors.

Symmetries of Playing Balls

A volleyball has type 3*2.

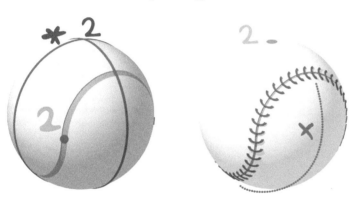

A tennis ball has mirror symmetry and rotations; its type is 2*2. A baseball has stitching, and its type is 2×.

A soccerball has type *532. What type does a basketball have? Find one and take a look!

Temari Balls

This ball has gyration points of order 2.

It also has a kaleidoscope ∗3; the type is 2∗3.

This ball has type 2 2 12.

Another example of an object with cubic symmetry, signature ∗432.

This ball has signature 532.

At first glance, this has type ∗22N for some large N. But if we pay very close attention to the weaving, the mirror symmetries are broken and the type is 22N.

Spherical Kaleidoscopes

Physical kaleidoscopes that generate spherical patterns are far less known than they should be. We attempt to remedy this by including plans for their construction! The balls on page 50 were made in just these sorts of kaleidoscopes.

The Di-scope, with signature *22N: Simply rest a pair of mirrors, meeting at an angle of $\frac{180°}{n}$ degrees, on a horizontal mirror. Can you explain the "mirrored corner paradox" that arises when $n = 2$: when peering into three mirrors meeting at a right-angle, your image is not reversed!

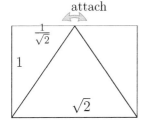
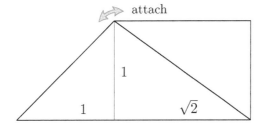

The Tetrascope and Octascope, with signatures *332 and *432: Cut mirrors as shown (above, right) and fold into a cone. Drop objects into the chamber for fascinating fun!

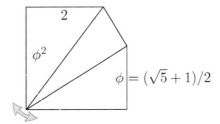

The Icosascope, with signature *532: This is the most marvelous of all! Cut the mirrors as shown above and fold into a cone. If you cut a hole on one end, along the gray lines, you will see a pattern in the shape of a stellated dodecahedron!

Bathsheba Grossman's Sculptures

Grossman's sculptures reveal our lack of full intuition about three dimensional symmetry; the symmetry type can really only be appreciated by holding the model and examining it from several points of view.

The sculpture *Quintrino* has a five-fold axis of rotational symmetry and so is easily recognized as having type 532.

We can verify this from other vantage points. Here we look down a two-fold axis of rotational symmetry.

We can guess that *Ora* has the rotational symmetry type 332 of the tetrahedron.

Here is a view of *Ora* down a two-fold axis.

The ones here are far more difficult to recognize.

Soliton is somewhat more mysterious. Here is a view down one 2-fold axis.

Here is another. There is one more two-fold axis—which we don't show here—and the type is 222.

Clef also has type 222, though it is hard to imagine from this one image how the views down the other axes might appear.

Can you guess what signature *Antipot* might have?

– 5 –

The Seven Types of
Frieze Patterns

There are other interesting patterns we've not yet considered. They are formed by the symmetries of plane patterns that repeat infinitely in one direction only: we call them *frieze patterns*. The facing page shows the seven different types of frieze pattern.

As we shall see in a moment, there is a Magic Theorem that we can use to list these. However, we don't really need it because any frieze pattern can be wrapped around a finite object such as a vase, which means that we can find all types of frieze patterns by looking at our results for the sphere.

According to the number of repetitions of the fundamental region, this vase will have one of the seven spherical symmetry types that involve a parameter N (namely NN, N×, N∗, ∗NN, 22N, ∗22N, or 2∗N), and so it's natural to say that the corresponding infinite frieze pattern has symmetry type ∞∞, ∞×, ∞∗, ∗∞∞, 22∞, ∗22∞, or 2∗∞. These could also be deduced from the following.

Theorem 5.1 (The Magic Theorem for Frieze Patterns) *The signatures of frieze patterns are precisely those that contain an ∞ symbol and cost exactly* $^\$2$.

The symbol ∞ costs $^\$1$, which makes perfect sense since $\frac{\infty-1}{\infty} = 1$. The symbol ∞ costs $^\$\frac{1}{2}$, since $\frac{\infty-1}{2\infty} = \frac{1}{2}$.

Figure 5.1 shows frieze patterns formed by footprints in the sand of an infinite desert plane. To analyze them, we transfer each one

You can find the signature for a frieze pattern just as in the Euclidean and spherical cases: imagine the pattern wrapped around the equator of a very big sphere.

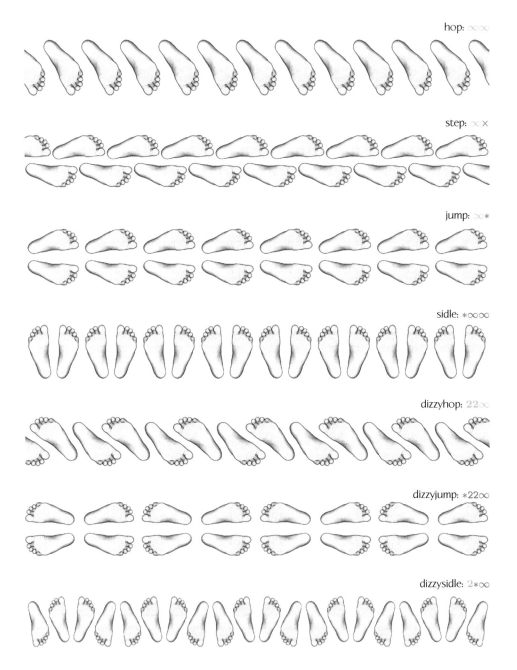

Figure 5.1. Tripping around the world in seven different ways!

to a finite spherical planet, where our previous methods show the resulting types to be NN, N×, N∗, ∗NN, 22N, ∗22N, and 2∗N for very large N. The originals were therefore ∞∞, ∞×, ∞∗, ∗∞∞∞, 22∞, ∗22∞, and 2∗∞, respectively.

The patterns for the types NN, N×, N∗, and ∗NN are what we get when we hop, step, jump, or sidle around the world. For the types 22N, ∗22N, and 2∗N, we spin between each hop, jump, or sidle, so we call these the "dizzy" types, or "ditypes."

With a little practice, the types can be found directly from the original patterns. For instance, the "dizzy jump" or "dijump" pattern has these mirror lines:

It is clear that the kaleidoscopes in this pattern have corners with angles of $\frac{\pi}{2}$ and that the kaleidoscopic points are 2-fold. We declare that the parallel sides of the kaleidoscope meet "at infinity" with an angle of $\frac{\pi}{\infty}$, and so the signature of this pattern is ∗22∞. In a similar way, the infinity symbols in the signatures of frieze patterns refer to translations (regarded as rotations about the infinitely distant poles).

Where Are We?

In Chapters 2–5, we've determined all possible types of symmetry for plane repeating patterns, spherical patterns, and frieze patterns using various forms of our Magic Theorem. So, the Magic Theorem is quite powerful. What we haven't done is explain why it is true! Because this theorem is so powerful you might think it would be hard, but the next chapter shows that, in fact, it's quite easy.

Exercises

1. What are the signatures of the friezes shown on page 66?

2. Take another look at the beautiful frieze patterns photographed by S. McBurney in Chicago (seen in Chapter 1) and analyze their types. Of course, more than one type appears in the top photograph!

3. What types are these alphabet friezes? Make up a few more!

ppppppp   bbbbbbbbb
                          ppppppppp
pdpdpdp

pqpqpqp   bdbdbdbdb
                          pqpqpqpqp

pbpbpbp   bdbdbdbdb
                          qpqpqpqpq

4. Check the types of these coffee friezes.

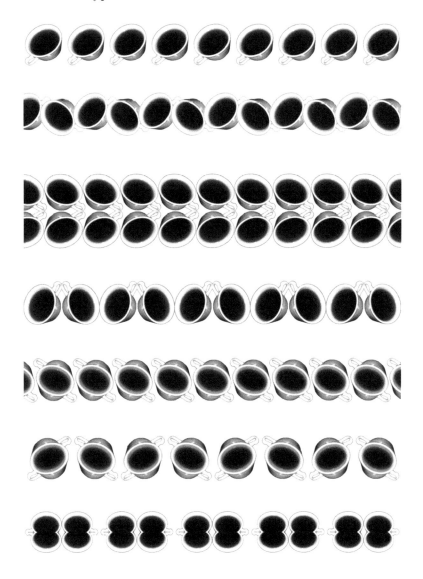

5. Analyze the appearences of this Sonny Bono look-alike. Be careful: some of these friezes have the same type and not every type is represented.

Answers to Exercises

1. Listed in order from top to bottom: $\infty\infty$, $\infty\times$, $\infty*$, $*\infty\infty$, 22∞, $*22\infty$, $2*\infty$.

2. Listed in order from top to bottom (including three in first photograph): 22∞, $*\infty\infty$, $\infty\times$, $\infty\infty$, $*\infty\infty$, $2*\infty$.

3. $\infty\infty$ $\infty*$

   22∞

           $*22\infty$

   $*\infty\infty$

      $\infty\times$ $2*\infty$

4. Listed in order from top to bottom: $\infty\infty$, $\infty\times$, $\infty*$, $*\infty\infty$, 22∞, $2*\infty$, $*22\infty$.

5. Listed in order from top to bottom: $\infty\infty$, $2*\infty$, 22∞, $\infty*$, $\infty\times$, $*22\infty$, $2*\infty$.

– 6 –

Why the Magic
Theorems Work

In this chapter we'll deduce the Magic Theorems from Euler's well-known theorem about maps. A mathematical map is like an ordinary map of countries and their borders.

We'll show how the different features of a symmetric pattern affect the structure of some specially chosen maps and how Euler's theorem is used to determine the costs assigned to the features of a signature. First, we'll consider the symmetries of finite objects, which, as we showed in Chapter 4, can be thought of as symmetries of the surface of a celestial sphere.

Folding Up Our Surface

We've told you that when several features are of the same kind you should count them only once. What this means is that we are really counting things not on the original surface but on a folded-up version of it, the folding taking all the points of the same kind to a single point. The set of points of the same kind is called the *orbit* of that kind of point under the action of the symmetry group, so this "orbit-folded" version of the surface is called the *orbifold*.

For example, the chair of Figure 6.1 has two symmetries: the trivial one and the reflection in its plane of symmetry. This reflection equates pairs of points in the left and right hemispheres, defining orbits. For example, the reflection equates the pair of blue points, and the pair of blue points is an orbit. The single red point lies on the mirror and is an orbit by itself.

(opposite page) A planar pattern lifted stereographically up to a sphere.

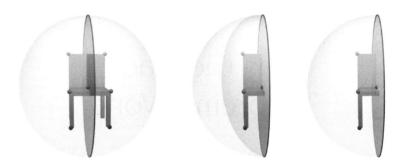

Figure 6.1. Folding a sphere (left) into a hemisphere (right). Matching points are fused to form an orbifold.

We can fold each orbit into a single point by pushing the right hemisphere into the left one as in the middle of Figure 6.1. The orbifold is therefore a hemisphere. Most points of the orbifold, like the blue, green, and yellow points, correspond to full-sized orbits (of two points), but the boundary of the orbifold consists of half-sized orbits like the red one. The signature for this pattern is $*$ and its cost is $\$1$. In the next section, we see how the cost of $\$1$ relates to the fact that the orbifold is a hemisphere.

Maps on the Sphere: Euler's Theorem

Leonhard Euler discovered a wonderful fact about maps drawn on a sphere—namely that $V - E + F = 2$, where V, E, and F are the numbers of vertices, edges, and faces of the map, respectively. We'll use *char* for $V - E + F$ since this number is traditionally called the *Euler characteristic*. The proof of Euler's Theorem is postponed to Chapter 7: for now we study what happens to *char* when we fold up our maps into orbifolds. On the left in the following figure, we see a map that has

$$V = 5, \qquad E = 8, \qquad F = 5.$$

So in accordance with Euler's Theorem, $char = 5 - 8 + 5 = 2$.

 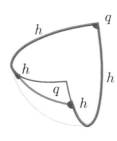

On the right we see the folded form of this map. Some of the vertices, edges, and faces have been halved (h) or quartered (q), so that we have

$$V = \tfrac{1}{2} + \tfrac{1}{2} + \tfrac{1}{4} = \tfrac{5}{4},$$
$$E = 1 + \tfrac{1}{2} + \tfrac{1}{2} = 2,$$
$$F = 1 + \tfrac{1}{4} = \tfrac{5}{4}.$$

So, for this quarter-spherical folded map, $char = V - E + F = \tfrac{5}{4} - 2 + \tfrac{5}{4} = \tfrac{1}{2}$.

In general, the same argument shows that under the folding corresponding to the symmetries of any spherical pattern, any map having the same symmetries as that pattern is taken to an orbifold map for which $char = 2/g$, where g is the number of symmetries of the pattern.

For example, when we fold a cubical map (shown below on the left) along the mirror lines indicated in the middle image below, we get a very simple orbifold map that has only $\tfrac{1}{6}$ of a vertex, $\tfrac{1}{4}$ of an edge, and $\tfrac{1}{8}$ of a face, so $char = \tfrac{1}{6} - \tfrac{1}{4} + \tfrac{1}{8} = \tfrac{1}{24} = \tfrac{2}{48}$. This is obvious because all we've done is take $\tfrac{1}{48}$ of $V - E + F = 8 - 12 + 6 = 2$ for the original cube map.

 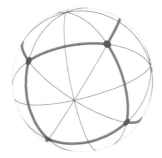 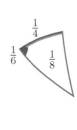

Why *char* = *ch*: Proving the Magic Theorem for the Sphere

We've now shown that for spherical types $char = 2/g$, so to prove the magic theorem in the spherical case we only need to explain why $char = ch$, the change after subtracting the cost of our signature from $^\$2$.

The following figures show how *char* changes as we add features to the orbifold. It is important to realize that in these arguments we can choose whichever map we like, since *char* doesn't depend on the map.

Punching a hole ($\ast$) decreases *char* by 1. Choose a map for which the "hole" is a single k-sided face. Then, removing it decreases F by 1 and V and E by $\frac{k}{2}$ (since vertices and edges around the hole get halved). Therefore, $V - E + F$ is reduced by $\frac{k}{2} - \frac{k}{2} + 1 = 1$.

Replacing an ordinary point by an N-fold cone point (N) decreases *char* by $\frac{N-1}{N}$. Choose a map for which the point is a vertex. Before the change, it contributes 1 to V; afterwards it contributes only $\frac{1}{N}$. The net change is

$$1 - \frac{1}{N} = \frac{N-1}{N}.$$

$\frac{1}{5}$

Replacing an ordinary boundary point by an N-fold corner point (N) decreases *char* by $\frac{N-1}{2N}$. Again, choose a map for which the given boundary point is a vertex. After the replacement, it will be $1/2N$ of a point, for a net change of $\frac{N-1}{2N}$.

The orbifolds for 13 of the 14 spherical signatures, namely

$\ast$532	$\ast$432	$\ast$332	$\ast$22N	$\ast$NN
		3$\ast$2	2$\ast$N	N$\ast$
532	432	332	22N	NN

can be obtained from the sphere (for which $char = 2$) by introducing holes, cone points, and corner points—the features symbolized by $\ast$, N, and N, respectively. The figures show that these changes to the

Figure 6.2. The orbifold of a pattern with signature × has *char* = 1. Bringing together opposite points on the sphere halves the sphere and so halves *char*.

Figure 6.3. Across the Euclidean plane in ∞ steps: around the world in 80 paces (or maybe less).

orbifold do indeed decrease *char* by 1, $\frac{N-1}{N}$, and $\frac{N-1}{2N}$, respectively. Since the costs of features in the signature equate with changes to the Euler characteristic, we find that *char* equals *ch*.

The fourteenth spherical signature is N×. The × stands for a *crosscap*, the orbifold obtained by folding each point of the sphere onto the point opposite it (Figure 6.2.) The end result is a weirdly twisted half sphere with *char* = 1. Crosscaps are discussed further in Chapter 8, where we'll also show that the list of 14 types is complete.

Otherwise, our Magic Theorem for sphere patterns is really just Euler's theorem $V - E + F = 2$ for maps on the sphere.

The Magic Theorem for Frieze Patterns

The Magic Theorem for frieze patterns is an easy consequence of the one for spherical patterns. This is because we can roll up an infinite frieze pattern into a finite one around the equator of a sphere, as in Figure 6.3.

The resulting spherical pattern will have a rotational symmetry of order N, and its symmetry will be one of the seven types ∗22N, 2∗N, 22N, ∗NN, N∗, N×, or NN, whose Euler characteristics have the form $\frac{1}{2N}$, $\frac{1}{2N}$, $\frac{1}{N}$, $\frac{1}{N}$, $\frac{1}{N}$, $\frac{1}{N}$, or $\frac{2}{N}$, respectively. The frieze pattern will correspondingly be one of $*22\infty$, $2*\infty$, 22∞, $*\infty\infty$, $\infty*$, $\infty\times$, or $\infty\infty$, whose Euler characteristics (obtained by letting N grow to ∞) are 0.

The Magic Theorem for Euclidean Plane Patterns

Here we must prove that any orbifold corresponding to a Euclidean plane pattern has Euler characteristic equal to 0. We do this by showing that, for any really large circular portion of the plane pattern, the Euler characteristic must be close to 0. In the proof, we use the fact that the numbers of vertices, edges, and faces inside the circular region is proportional to the area of the circle and so to the square of its radius, while the numbers of vertices, edges, and faces along the boundary of the region are just proportional to the length of the boundary and to the radius of the circle.

To begin the proof, take a map having the same symmetry as the pattern and delete everything that lies outside a circle of large radius R on it. Wrap the circular patch P of the map around a large sphere; this turns a region of our planar map into a map on the sphere.

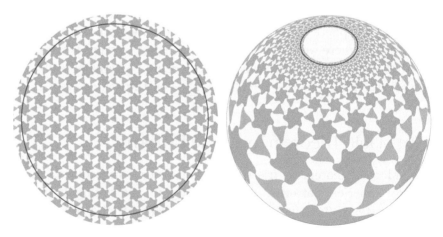

The numbers V, E, and F for the portion P of the infinite map will be close to Nv, Ne, and Nf, where v, e, and f are the (possibly fractional) numbers of these things on the orbifold of the original map and where N is the number of copies of this orbifold completely covered by the portion P. Since the area of P is just πR^2, this number N will be approximately kR^2 for some positive number k.

In fact, the differences $V - Nv$, $E - Ne$, and $F - Nf$ between the actual V, E, and F and their approximations will be bounded by multiples of R. This is because the "extra" vertices, edges, and faces belong to copies of the fundamental region of the map that lie across the perimeter of P. The perimeter has length $2\pi R$, so the number of copies of the fundamental region that overlap the perimeter is proportional to R.

We can therefore suppose that

$$(V - Nv) - (E - Ne) + (F - Nf) < cR$$

for some number c, and so

$$ch = v - e + f < |(V - E + F + cR)/N| < |(2 + cR)/kR^2|.$$

Since the right-hand side of this inequality tends to zero as R tends to infinity, it must be true that $ch = 0$.

Where Are We?

We have just shown that plane patterns always have $char = ch = 0$, and so we have completed our justification of the costs given for the symbols in the signature by proving the Magic Theorem for the plane.

Up to now it has been important to distinguish between the red and blue digits in our signatures because they have different costs. From now on we'll feel free to print them in black. It's easy to recover the proper colors, if you want them; the symbols that should be blue are just those before the first cross ($\times$) or star ($*$).

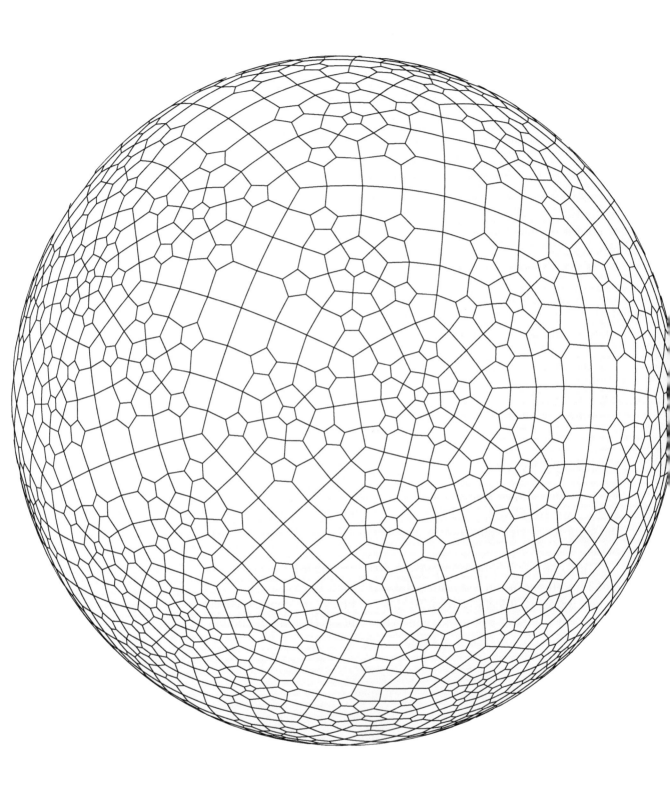

– 7 –

Euler's Map Theorem

We've made some powerful deductions from Euler's Theorem that $V - E + F = 2$ for maps on the sphere. Now we'll prove it!

Proof of Euler's Theorem

We can copy any map on the sphere into the plane by making one of the faces very big, so that it covers most of the sphere.

This is really quite familiar—we've all seen maps of the Earth in the plane.

We'll think of this big face as the ocean, the vertices as towns (the largest being Rome), the edges as dykes or roads, and ourselves as barbarian sea-raiders! (See Figure 7.1.)

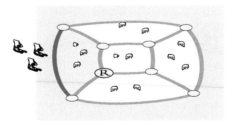

Figure 7.1. Our prey.

(opposite page) Like all maps on the sphere, this beautiful map (signature **532**) has $V + F - E = 2$.

In this new-found role, our first aim is to flood all the faces as efficiently as possible. To do this, we repeatedly break dykes that separate currently dry faces from the water and flood those faces. This removes just $F - 1$ edges, one for each face other than the ocean, by breaking $F - 1$ dykes.

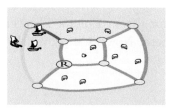
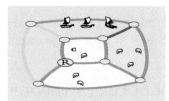

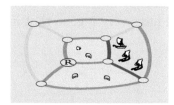

Deleting an edge decreases the number of edges by 1 and also decreases the number of faces by 1, so $V - E + F$ is unchanged.

We next repeatedly seek out towns other than Rome that are connected to the rest by just one road, sack those towns, and destroy those roads. (See Figure 7.2.)

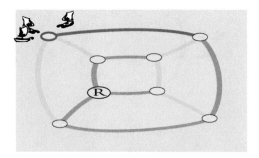

Figure 7.2. Our raid continues!

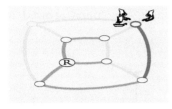

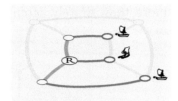

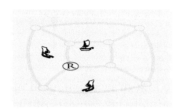

Deleting a vertex and the edge joining it to the rest of the tree does not change $V - E + F$ either, so $V - E + F$ must always have been 2.

We have sacked $V - 1$ towns by destroying $V - 1$ roads, one for each town other than Rome. The number of edges in the original map must therefore have been $(F - 1) + (V - 1) = V + F - 2 = E$. Therefore, $V + F - E = 2$, proving Euler's Theorem.

(Did we sack every town other than Rome? Yes; an unsacked town furthest from Rome would have two paths back to Rome, which however must enclose some dry fields, a contradiction. Did we destroy all remaining roads? Yes; an undestroyed road must be between unsacked towns, which must both be Rome; but then again it must enclose some dry fields.)

We have tacitly assumed that each face is a topological disk, and we will continue to suppose this. We have also taken for granted some intuitively obvious facts about the topology of the sphere whose formal proofs are surprisingly difficult.

The number 2 is Euler's characteristic number for the sphere. Every surface has such a number.

The Euler Characteristic of a Surface

Theorem 7.1 *Any two maps on the same surface have the same value of* $V - E + F$, *which is called the* Euler characteristic *for that surface.*

We prove that any two maps on the same surface have the same Euler characteristic $V - E + F$ by considering a larger map obtained by drawing them both together. We shall suppose that no two edges meet more than finitely often, pushing the maps around a bit if necessary.

We first draw one map in black ink, the other in red pencil. Then we gradually ink in parts of the pencil map, noticing that $V - E + F$ does not change. The following figures show the first few steps of this process for a pair of maps.

Inserting a vertex. V increases by 1, E increases by $2 - 1 = 1$, so $V - E + F$ increases by $1 - 1 + 0 = 0$.

Inserting an edge. E increases by 1, F increases by $2 - 1 = 1$, so $V - E + F$ increases by $0 - 1 + 1 = 0$.

We can continue to make these insertions, gradually inking in the entire figure and not changing $V - E + F$:

This argument shows that the characteristic number $V - E + F$ for the compound map is the same as that for the originally black map. Equally, it's the same for the originally red map! Therefore, those two original maps must have had the same characteristic.

The Euler Characteristics of Familiar Surfaces

Let us work out a few examples.

The Euler characteristic of a torus is 0.

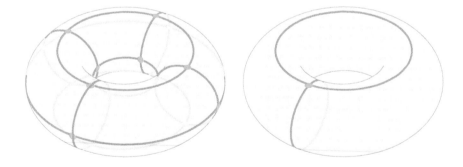

The map on the left has 16 vertices, 32 edges, and 16 faces, so $V - E + F = 16 - 32 + 16 = 0$. The map on the right is much simpler: it has just 1 vertex, 2 edges, and 1 face, so $V - E + F = 1 - 2 + 1 = 0$. The theorem tells us that we can use either map to work out the characteristic.

The Euler characteristic of an annulus or Möbius band is 0.

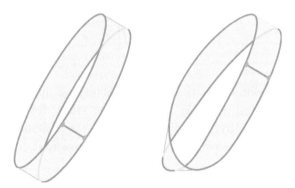

On the left, we see a map on an annulus, on the right a map on a Möbius band. Both maps have 2 vertices, 3 edges, and 1 face, and so $V - E + F = 0$.

The Klein bottle also has Euler characteristic 0.

The Klein bottle, a one-sided, boundary-less surface, also has Euler characteristic 0. Again, we choose a map with just 1 vertex, 2 edges, and 1 face, yielding $V - E + F = 1 - 2 + 1 = 0$.

A sphere with n holes punched in it has Euler characteristic $2 - n$.

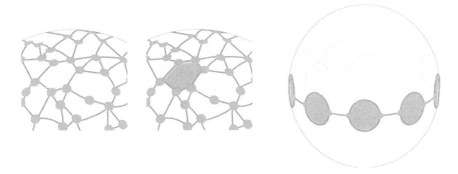

We may see this easily by taking a map on the sphere that has a great many more than n faces. If we delete n non-adjacent faces, we have kept V and E the same but decreased F by n. Consequently, the Euler characteristic will be n less than that of a sphere: $2 - n$. (In fact, punching n holes in any surface will always decrease the Euler characteristic by n.) Alternatively, we may systematically design a map specifically for this surface. On the right above, we see a map with $2n$ vertices, $3n$ edges, and 2 faces, and so $V - E + F = 2n - 3n + 2 = 2 - n$. As we will see in Chapter 8, a disk is topologically equivalent to a sphere with one hole in it, and so a disk has Euler characteristic 1.

An n-fold torus has Euler characteristic $2 - 2n$.

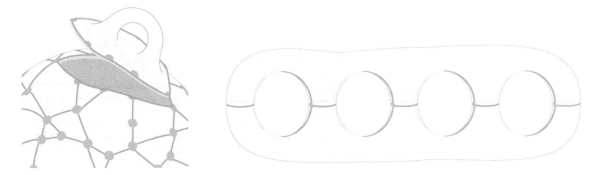

An n-fold torus is a surface obtained from a sphere by adding n handles, or equivalently n tunnels. We make it by deleting n faces from a sphere and then attaching n handles. Each handle is just a torus with a (very large) hole punched in it and will contribute $0 - 1$ to the total Euler characteristic. Each hole punched in the sphere will contribute -1. So the net result is that the Euler characteristic of an n-holed torus is $2 - 2n$. Or we may design a map specifically for this surface, with $2n$ vertices, $4n$ edges, and 2 faces: $V - E + F = 2n - 4n + 2 = 2 - 2n$.

Two mystery surfaces with Euler characteristic –2.

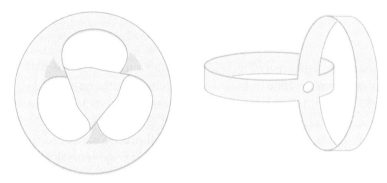

Here we have two mystery surfaces with $V - E + F = -2$. Both have two boundaries and are two-sided; in Chapter 8, we will learn that they then must be the same surface, topologically. In the meantime, you might try to decide for yourself whether this is obvious!

Where Are We?

In this chapter we have shown that for the sphere the Euler characteristic is 2 and more generally that the value of $V - E + F$ depends only on the surface on which a map is drawn and not on the map itself. This supports the proof of the Magic Theorem in Chapter 6, which in turn supports the enumeration of symmetry types in Chapters 2–5.

In the next chapter we shall classify all possible surfaces, which will show us all the forms an orbifold could possibly take and will help us conclude that we've enumerated the signatures of all possible symmetry types.

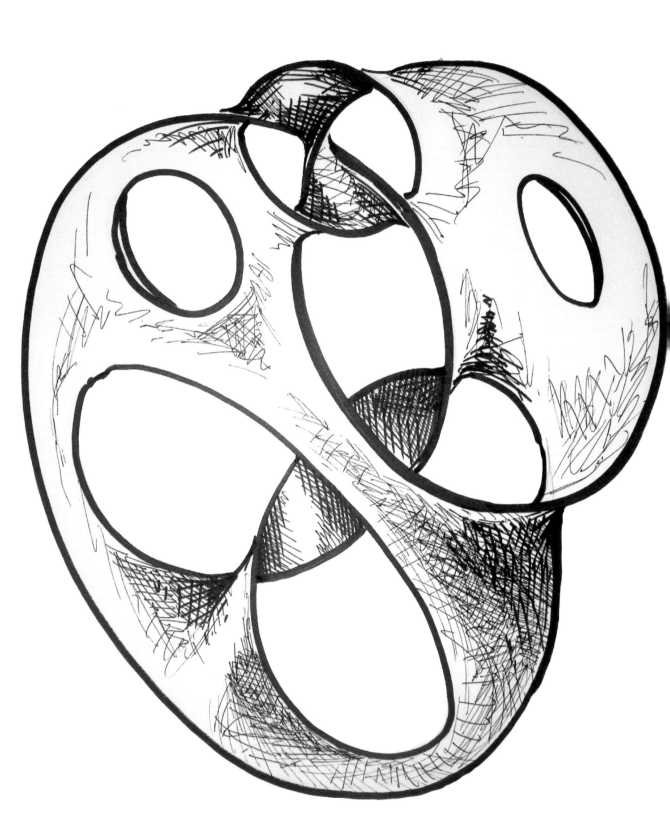

– 8 –

Classification of Surfaces

In Chapters 2–5, we gave a supposedly complete list of symmetry types of repeating patterns on the plane and sphere. Chapters 6–7 justified our method of "counting the cost" of a signature, but we have yet to show that the given signatures are the only possible ones and that the four features we described are the correct features for which to look.

Any repeating pattern can be folded into an orbifold on some surface. So to prove that our list of possible orbifolds is complete, we only have to show that we've considered all possible surfaces.

In this chapter we see that any surface can be obtained from a collection of spheres by punching holes that introduce boundaries (∗) and then adding handles (○) or crosscaps (×). Since all possible surfaces can be described in this way, we can conclude that all possible orbifolds are obtainable by adding corner points to their boundaries and cone points to their interiors. This will include not only the orbifolds for the spherical and Euclidean patterns we have already considered, but also those for patterns in the hyperbolic plane that we shall consider in Chapter 17.

Caps, Crosscaps, Handles, and Cross-Handles

Surfaces are often described by identifying some edges of simpler ones. We'll speak of zipping up zippers. Mathematically, a zipper ("zip-pair") is a pair of directed edges (these we call *zips*) that we intend to identify. We'll indicate a pair of such edges with matching arrows:

(opposite page) This surface, like all others, is built out of just a few different kinds of pieces—boundaries, handles, and crosscaps. But it may be hard to tell how, at just a glance!

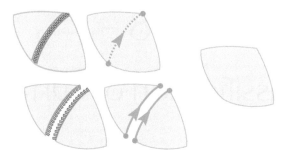

There are simple modifications that you can make to a surface by zipping together the boundaries of one or two holes. If a single hole is bounded by a clockwise zip and its counterclockwise mate, we have a *cap*: zipping this up just seals the hole, so we can ignore it.

Zipping a cap.

If instead the two zips are in the same sense (e.g., both counterclockwise), we have the instructions for what's called a *crosscap*. To get a clear picture is rather difficult: the usual one involves letting the surface cross itself along a line, leading to an 8-shaped cross-section as shown in the lower figure.

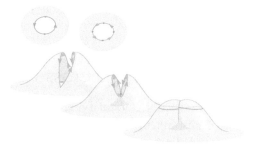

Zipping a crosscap.

We start by dividing each zip into two zips, as in the top of the figure. Then we distort the surface, bringing the two sets of zips together. We obtain something like the final surface.

If two nearby holes on a surface are bounded by zips in opposite senses, we have the instructions for a *handle*. To see this, let the two "tubes" grow out of the same side of the surface and then meet.

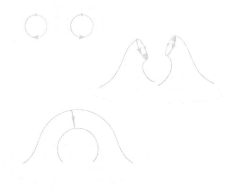

Zipping a handle.

If such zips are in the same sense, we can let the "tubes" grow out of opposite sides of the surface to form a *cross-handle*, which is sometimes called a *Klein handle*.

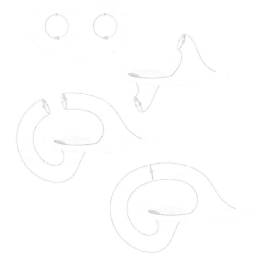

Zipping a cross-handle.

Any one sphere with the instructions for three handles, two cross-caps, and two holes, for example, is topologically the same as any other sphere with the instructions for three handles, two crosscaps, and two holes, since we can just push the holes around. So, the important point is just how many of each of these things there are for each component.

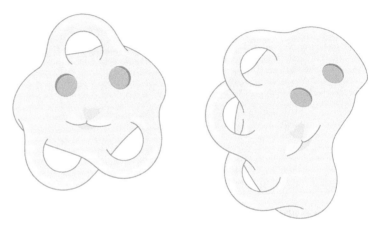

Two equivalent, tidy surfaces. Each has three handles, two holes, and one crosscap.

Lemma 8.1 (Tidying Lemma) *Every surface is topologically equivalent to a "tidy" one, obtained from a collection of spheres by adding handles ($\circ$), holes ($*$), crosscaps ($\times$), and cross-handles ($\otimes$).*

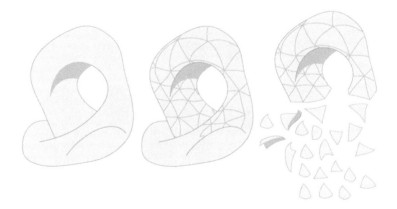

We will suppose that the surface is given to us as a collection of triangles that have zips indicating how they should be pieced together. (In technical language, this is called a "triangulable 2-manifold." It is a deep and difficult theorem, proved by Tibor Rado in 1925, that every compact 2-manifold is triangulable.)

Figure 8.1. A triangle is a tidy surface—it's a sphere with a hole in it.

Since topologically a triangle is just a sphere with a hole (Figure 8.1), it's certainly tidy before we do any zipping up. So, all we need to prove is that we can zip up any one zip-pair of a tidy surface in such a way as to preserve its tidiness.

The lemma is obvious in the "snug" cases when the two zips of this zipper together occupy all the boundary components they involve (Figures 8.2–8.4). But Figures 8.5–8.9 show that it is almost as obvious in the "gaping" cases when they don't, since these produce the same surfaces as the snug ones, with an extra boundary or two. (The figures illustrate only the "totally gaping" cases.)

In fact, we can improve the lemma.

Theorem 8.2 (The Classification Theorem for Surfaces) *To obtain an arbitrary connected surface from a sphere, it suffices to add either handles or crosscaps and maybe to punch some holes, giving boundaries. So, the symbols $\circ^a *^b$ and $*^b \times^c$ represent all possible surfaces.*

See the next two sections to find out why.

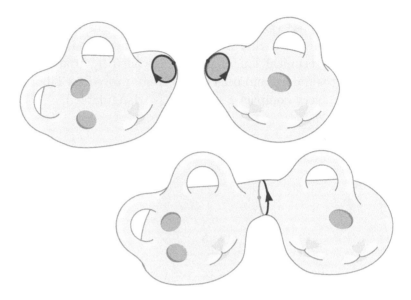

Figure 8.2. Zips on different components of a surface. From $*^{a+1} \circ^{b} \times^{c} \otimes^{d}$ and $*^{A+1} \circ^{B} \times^{C} \otimes^{D}$, we get $*^{a+A} \circ^{b+B} \times^{c+C} \otimes^{d+D}$.

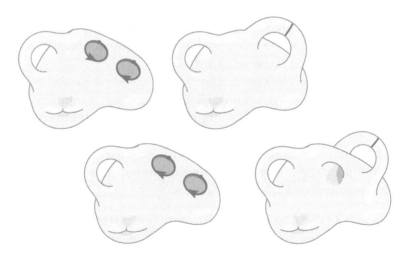

Figure 8.3. Zips on different boundaries of the same surface component. At top we zip a pair with opposite orientations. On the bottom we zip a pair with the same orientation. From $*^{a+2} \circ^{b} \times^{c} \otimes^{d}$, we get $*^{a} \circ^{b+1} \times^{c} \otimes^{d}$ or $*^{a} \circ^{b} \times^{c} \otimes^{d+1}$ according to the orientations of the zips.

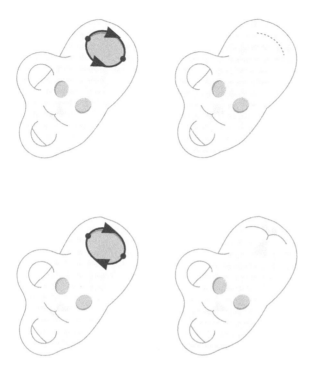

Figure 8.4. Zips on same boundary. From $*^{a+1} \circ^b \times^c \otimes^d$, we get $*^a \circ^b \times^c \otimes^d$ or $*^a \circ^b \times^{c+1} \otimes^d$ according to the orientations of the zips.

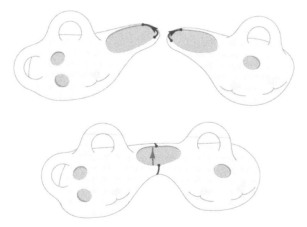

Figure 8.5. Gaping zips on different components form a joined surface with boundary.

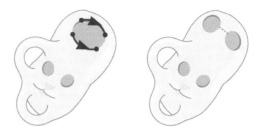

Figure 8.6. Gaping zips with opposite orientations on the same boundary form a cap with boundaries. From $*^a \circ^b \times^c \otimes^d$, we obtain $*^{a+1} \circ^b \times^c \otimes^d$.

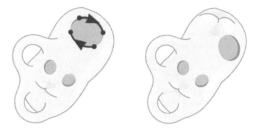

Figure 8.7. Gaping zips with the same orientation on the same boundary form a crosscap with boundaries. From $*^a \circ^b \times^c \otimes^d$, we obtain $*^a \circ^b \times^{c+1} \otimes^d$.

Figure 8.8. Gaping zips with the same orientation on different boundaries of the same component form a crosshandle with a boundary. From $*^{a+1} \circ^b \times^c \otimes^d$, we obtain $*^a \circ^b \times^c \otimes^{d+1}$.

Figure 8.9. Gaping zips with opposite orientations on different boundaries of the same surface form a handle with boundaries. From $*^{a+1} \circ^b \times^c \otimes^d$, we obtain $*^a \circ^{b+1} \times^c \otimes^d$.

We Don't Need Cross-Handles

This is shown by zipping up Figure 8.10(b) in two ways. If we do up the blue horizontal zipper first, we get the instructions for a cross-handle ($\otimes$) (Figure 8.10(a)); therefore, doing up both zippers will give a cross-handle.

Alternatively, we can find what would result (Figure 8.10(c)) from doing up the green vertical zipper first from the general theory. Namely, we get the crosscap ($\times$) that would come from the corresponding "snug" case, together with a boundary formed by the blue zips. But this boundary is just the instructions for another crosscap, so what we've proved may be expressed by an equation:

$\otimes = \times\times$; a cross-handle may be replaced by two cross-caps.

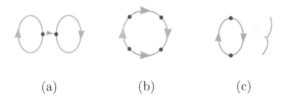

(a)   (b)   (c)

Figure 8.10. A cross-handle (a) is just a combination of the two crosscaps (c), since they are both generated by the same set of instructions (b).

We Don't Need to Mix Crosscaps with Handles

If we have both a crosscap ($\times$) and the instructions for a handle ($\circ$), we can take the hole at one end of the handle for a "walk" through

Figure 8.11. By moving one of the holes to be zipped along the surface (and through the crosscap) in the right way, the instructions for a handle become the instructions for a cross-handle.

the crosscap so that it returns with the reversed orientation. We will then have instructions for a cross-handle (Figure 8.11). Symbolically, this proves that $\circ\times$ can be replaced by $\times\times\times$ or $\times^3$. More generally, $\circ^a\times^b = \times^{\,2a+b}$ if $b > 0$.

That's All, Folks!

We cannot simplify this system for describing surfaces any further beause all these surfaces are topologically distinct. This is because

- $\circ^a *^b$ is orientable, with b boundary components and Euler characteristic $= 2 - b - 2a$,

while

- $*^b \times^c$ ($c > 0$) is non-orientable, with b boundary components and Euler characteristic $2 - b - c$,

so that the numbers a, b, and c are invariants.

In particular, we can use one of $\circ^a *^b$ and $*^b \times^c$ to indicate the topological type of an orbifold. But, an orbifold differs from an abstract surface just because it has local features coming from points that were fixed by some symmetries. Since we showed in Chapter 1 that the only possibilities for the symmetries fixing a point are **n** and $*$**n**, there can be no other local features than gyration points and kaleidoscopic points.

This proves at last that the four fundamental features that make up our signature symbol

wonders	gyrations	kaleidoscopes	miracles
$\circ...\circ$	**AB...C**	$*$**ab...c**$*$**de...f**$...$	$\times...\times$

really are all that's needed to specify its orbifold. In turn, this finishes our discussion of planar and spherical groups, since we saw in Chapters 3–5 that these are determined up to isotopic reshaping by their orbifolds.[1]

The miracles and wonders in Chapter 2 were just a poor man's way of approaching the global topology of the orbifold surface. We

---

[1] For hyperbolic groups, there is something more to say, as we shall see in Chapter 17.

Euler Characteristics of Standard Surfaces

In the last chapter, we showed that the Euler characteristic of a given surface was independent of its triangulation. To work out the Euler characteristic, we can make our triangulations as nice as we please. The following figures show that

- punching a hole (✶) decreases the Euler characteristic by 1,

    as does

- adding a cross-cap (×),

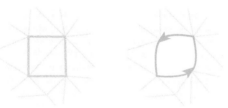

    while

- adding a handle (◯) decreases the Euler characteristic by 2.

We know that a sphere has Euler characteristic 2; hence, the Euler characteristic is

$$2 - 2a - b \text{ for } \bigcirc^a {*}^b,$$

$$2 - b - c \text{ for } {*}^b {\times}^c.$$

can now formally define them by saying that a pattern "has just a wonders" or "has just c miracles" according as this surface is an orientable one, $\circ^a *^b$, or a non-orientable one, $*^b \times^c$.

Where Are We?

Chapter 1 showed that local symmetries must be kaleidoscopic or gyrational, and in Chapter 2 we added miracles and wonders to obtain our four fundamental features. Supposing that these were enough, we then enumerated the symmetry types of planar and spherical patterns in Chapters 3, 4, and 5, using the Magic Theorem that Chapter 6 deduced from Euler's Theorem, proved in Chapter 7.

In this chapter we have proved the Classification Theorem for Surfaces, which shows that miracles and wonders (now properly defined as crosscaps and handles) can describe the global topology of any orbifold (see also [30]). Putting everything together, this shows that our four fundamental features suffice for the entire structure, so completing the investigation, we therefore state that our lists of Euclidean and spherical groups are indeed complete.

Is this all? No! So far we've mentioned only the Euclidean and spherical signatures, which cost at most $^\$2$. But we've really classified the more expensive ones too, and we'll see some of the lovely patterns to which they correspond in Chapter 17.

Examples

We have shown that every surface is topologically a sphere, possibly with some number of holes, possibly with some number of crosscaps or handles. But what should we make of the following strange surface?

Our classification theorem is actually an algorithm, a process, for breaking apart our surface and then putting it back together again, in a tidy form. In practice, though, this can be a little tedious.

We can immediately work out the tidy form of a surface just by drawing a map on the surface and calculating its Euler characteristic, counting the number of boundaries, and checking whether the surface is orientable.

Our strange surface has Euler characteristic −2, has two boundaries (which we can check by tracing a finger around each boundary curve), and is orientable. The topology of this surface is not visible at a glance but can only be ○∗∗—the surface is a torus with two holes (a *twice-punched* torus).

What are the tidy forms of familiar surfaces?

Punched or Punctured?

We distinguish between *punching* a surface, by removing a disc, and *puncturing* it, by removing a single point. Most topologists use "puncturing" in both senses.

The disk, of course, has signature ∗; it is an orientable surface with Euler characteristic 1 and one boundary.

The annulus has Euler characteristic 0, is orientable, and has two boundaries; it is therefore ∗∗—a twice-punched sphere.

The torus has Euler characteristic 0, is orientable, and has no boundary; it is— unsurprisingly!—○, a sphere with a handle.

The Möbius band has Euler characteristic 0, is non-orientable, and has one boundary; it is ∗×, a punched crosscap.

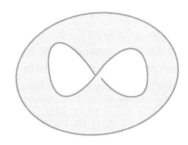

The Klein bottle has Euler characteristic 0, is non-orientable, and has no boundary; it is ×× , a pair of crosscaps.

This strange-looking surface is just a punched Möbius band, ∗∗×: it has Euler characteristic −1, has two boundaries, and is non-orientable.

– 9 –

Orbifolds

In Chapter 6 we showed how to fold up a pattern into a surface that we called its orbifold. We didn't really discuss the peculiar surfaces we get, because we didn't need to, but maybe you'll be interested to see some of them.

In this chapter, we give a number of examples of patterns on surfaces, which you can use to increase your understanding of orbifolds. There is very little text: we prefer to explain things largely by picture.

You might even want to make your own orbifolds. We suggest you buy a spare copy of the book before cutting out the patterns on the next few pages, so as to increase our royalties. Or if you don't have enough money for that, feel free to photocopy them instead. Can you fold or roll each pattern up so that all the faces coincide?

(opposite page) Each type of symmetry has its orbifold, found by folding or rolling the pattern up so that corresponding points are brought together. For example, to find the orbifold of this wallpaper pattern with signature ∗∗, we first fold the pattern along its mirror lines, into a fan shape and then into a strip. We then roll this strip into a cylinder—the orbifold of the pattern. All curlicues have been brought together to coincide on the final surface.

A kaleidoscopic point.

A gyration point.

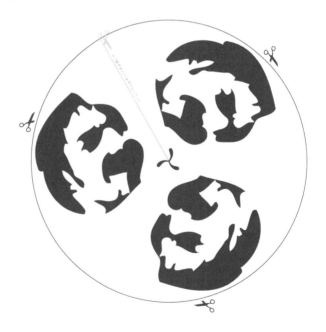

Three frieze patterns.

Here are the orbifolds for the previous patterns:

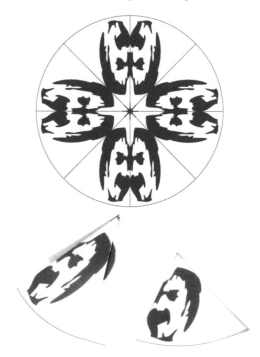

The signature is ***4**. A kaleidoscopic corner is folded up into just that—a corner of a kaleidoscope!

The signature is **3·**. A gyration point is rolled into a cone.

The signature is ∗∞∞. A pair of reflections folds fan-like into a strip.

The signature is ∞∞. The orbifold is a cylinder.

The signature is ∞×. The orbifold is a Möbius band.

Now try this for some of the other patterns in the book.

The orbifold for **3∗3** is a topological disk, with a cone point of order 3 in the interior and a corner point of order 3 on its mirror boundary.

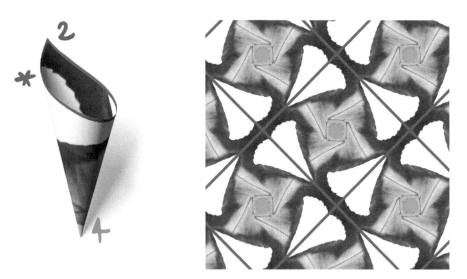

The orbifold for **4∗2** is also a topological disk, with one cone point and one corner point.

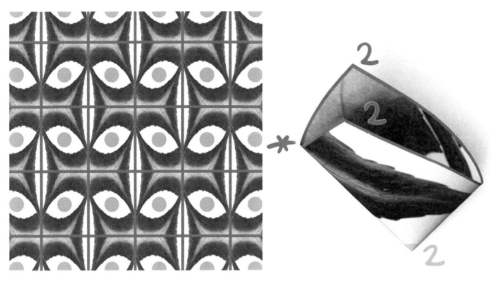

The orbifold for **2∗22** is a topological disk, with one cone point and two corner points.

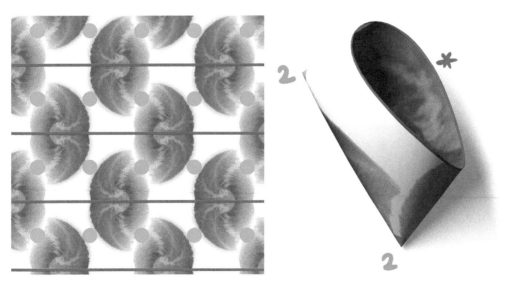

The orbifold for **22**∗ is a topological disk, with two cone points. The "kaleidoscope" here, of type ∗, has no corner points.

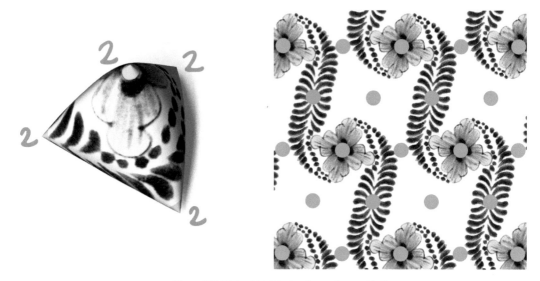

The orbifold for **2222** is simply a sphere with four cone points.

Part II

Color Symmetry, Group Theory, and Tilings

Introduction to Part II

The first part of this book provided the details needed to classify the symmetries of repeating patterns on the plane and sphere using the method of orbifolds and the signature notation for them. Traditionally, this enumeration was done using group theory. In the central section of the book, we show how our method relates to the traditional one, by way of a discussion of color symmetries. We continue to provide examples and illustrations to accompany the material.

So far we've avoided group theory, except for an aside in Chapter 4 but we expect readers of Part II to know some group theory. We expect such readers will be surprised that we classified these groups without using any group theory! Surprisingly, group theory doesn't really help.

In the upcoming chapters we once again place the results before the theorems, this time discussing color symmetries using the ideas supported by the discussion in Chapter 14.

The new method we used for classifying repeating patterns was pioneered by Murray McBeath and is now recognized as part of the wide-ranging theory of groups and manifolds that William Thurston has made his own. Our signature is a shortened form of McBeath's, which conveys exactly the same information.

We have been following what one can call Thurston's commandment:

> Thou shalt know no geometrical group save by understanding its orbifold.

Our orbifold signature is only one—which we hope is the last one—of several systems of names for these groups. Dictionaries between it and the other systems are given in Tables A.1 and A.2 of the appendix. There is an important way in which it differs from all the other systems except McBeath's; namely, these systems usually name groups by somehow specifying their generators. After mathematicians have been thinking for a century, we can see that this was not a good way to proceed—after all, it disobeys Thurston's commandment! Because groups have many generating sets, each author has made a more or less arbitrary choice of which generators to use and how to indicate them. Moreover, since their systems are usually not theoretically complete, difficult cases must often be distinguished by adding arbitrary signs.

It is important to realize that the characters in the signature do *not* correspond to generators in this way. What they do name (in compliance with the commandment!) is features of the orbifold, which may correspond to local singularities (gyration points and kaleidoscopes) or global properties of the topology of the orbifold (wonders and miracles).

There is a vague relationship between these notions and the group elements that are responsible for them, but it's far from being a one-to-one correspondence and is usually misleading. For instance, there are two **2**'s in our name **22** for a certain group despite the fact that that group has just one order-2 rotation. Why? Because that rotation creates two singularities on the orbifold. Again, the two ∗'s in ∗∗ don't really correspond to a generating set. Each ∗ is caused by a reflection, but those two reflections do not suffice to generate the group!

However, there must be a way to recover group generators from the signature since everything about the group can be read from its orbifold. We provide it in the next chapter, which sets forth the rules for obtaining group presentations from signatures. The following chapters then apply these presentations to the problem of finding symmetric "colorings" of repeating patterns.

The complete lists of all primefold colorings of repeating patterns in the plane and sphere appear for the first time in this book. You can understand these lists without needing to follow the technical arguments of Chapters 11–13 that deduce them from the group presentations. Following our practice, group presentations are finally justified in Chapter 14.

We turn to two other subjects in the last two chapters of Part II. Chapter 15 uses the orbifold idea to enumerate interesting tilings, while Chapter 16, as a digression, enumerates the abstract groups.

Some Group Theory

In this next portion of the book we will be discussing the groups of symmetries of our patterns. Here we give a gentle introduction to mathematical group theory.

In Chapter 4 we said that the symmetries of a pattern form a group. This means something quite specific. Consider this gyroscopic pattern.

It has exactly twelve symmetries: we may rotate by 30°, 60°, ... on up to 330°. That makes eleven, and the twelfth symmetry—the *identity*—is the one that does nothing at all or, equivalently, rotates through 360°.

These symmetries may be combined: we may first rotate by, say, 240° and then by 150°; the end result would be the same as rotating by 390° or, more simply, 30°, which (of course) is also one of our twelve symmetries.

The identity is special: when we combine the identity with any other symmetry, we don't change the result. Also, every symmetry has an *inverse* that is its undoing; combining a symmetry with its inverse produces the identity.

In general, a pair of symmetries A and B of a pattern have a product AB, obtained by performing the motions of the pattern corresponding to A and B one after the other, producing another symmetry in the group. This production is *associative*: performing motion AB followed by motion C is the same as performing motion A followed by motion BC. There is an identity symmetry, 1, which doesn't move the pattern at all; in particular, for any symmetry A, $A1 = A$ and $1A = A$. Finally, for every symmetry A there is an inverse symmetry A^{-1} with $AA^{-1} = 1$ and $A^{-1}A = 1$. These conditions are precisely those that define a mathematical *group*. How much information is needed to specify a group? In the example above, all the symmetries could be formed by applying just one symmetry repeatedly—say, rotating by $30°$—over and over again. We say that this symmetry, which we'll call α, *generates* the group.

Moreover, some symmetries can be made in many ways; for example, the rotation by $60°$ can be achieved by applying the rotation α twice, fourteen times, or twenty-six times. But knowing that $1 = \alpha^{12}$ suffices to explain this, and we call this equation a *relation* of the group.

A *group presentation* consists of a list of generators and relations that suffice to describe the group. The figure's group of gyroscopic symmetries is fully described by knowing it has one generator α and one relation $1 = \alpha^{12}$. We can summarize the presentation by writing $\langle \alpha \mid 1 = \alpha^{12} \rangle$.

In this next example, of a pattern with kaleidoscopic symmetry, there are ten symmetries: five are reflections across the marked mirror lines, four are rotations of $72°, 144°, 216°,$ and $288°$, and the tenth is the identity. When we combine these, do we really get a symmetry on our list? If we reflect across line P and then Q, the deep blue lobe is first taken to the medium blue one and then on to the light blue one. The net effect is that we rotated counterclockwise by $72°$. Try combining other symmetries yourself! In general, applying first one and then another of the symmetries of any pattern will result in a symmetry. But, the order in which these symmetries are applied can make a difference! PQ amounts to a counterclockwise rotation but QP is clockwise.

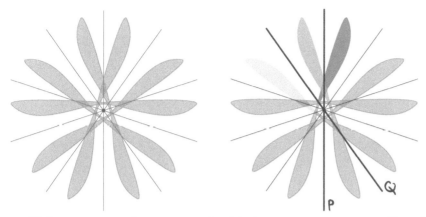

   This group can be presented with just two generators, P and Q—all the symmetries are formed from these—and three relations, $1 = P^2, 1 = Q^2$, and $1 = (PQ)^5$, that explain all other equivalences. We can summarize this presentation by writing $\langle P, Q \mid 1 = P^2 = Q^2 = (PQ)^5 \rangle$.

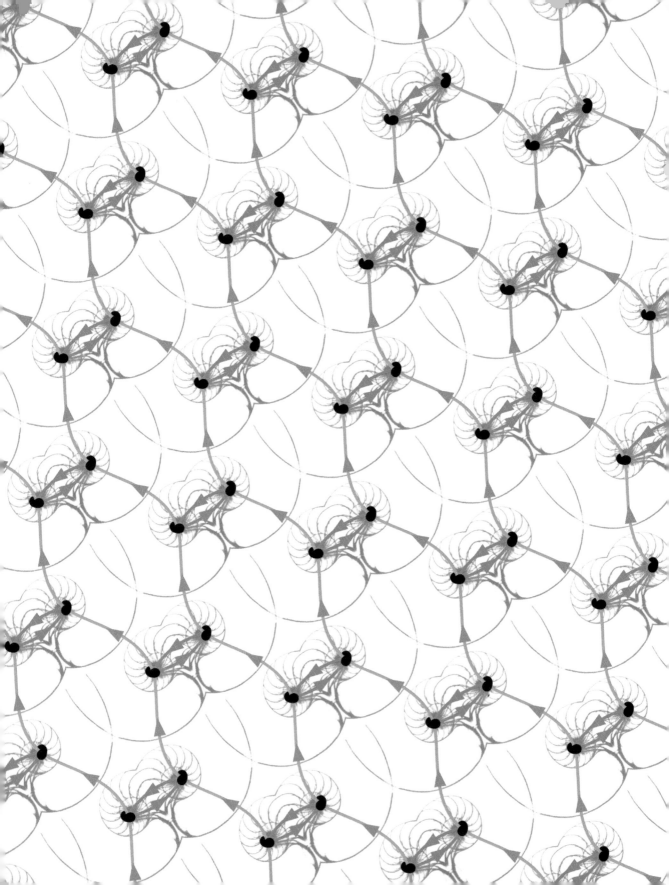

– 10 –

Presenting Presentations

In later chapters we will use some simple ideas from group theory to enumerate symmetric colorings of the patterns discussed in the first third of this book. In this chapter we tell you how to find generators and relations for the symmetry group of a pattern from its signature. The proof that the resulting presentations are correct is (of course!) postponed until Chapter 14.

Generators Corresponding to Features

There are some generators for each feature and also some relations arising from that feature. In addition to these *local relations*, there is a single *global relation*, which asserts that the product of the "Greek" generators, of which there is one per feature, is trivial. We indicate the generators for the various features by *annotating* the signature in a manner that we shall now explain.

For a *handle* ($\circ$), there is a Greek generator, say α, and two "Latin" ones, say X and Y, subject to the local relation

$$X^{-1}Y^{-1}XY = \alpha,$$

for which the annotation is

$$\alpha_{\circ}{}^{X,Y}.$$

For a *gyration point* ($\mathbf{A}$) of order A, we have only a single Greek generator, say β, and a single relation,

$$1 = \beta^{A},$$

(opposite page) The presentation for $*\times$. The orange "lenses" are the relator P^2, and the green "vases" are the relator PZ^2PZ^{-2}.

annotated by

$$^{\beta}\mathbf{A}.$$

For a *kaleidoscope* ($*\mathbf{ab...c}$) with n types of corner point, there is a Greek generator, γ, together with $n+1$ Latin ones, $P, Q, ..., T$, subject to the relations

$$1 = P^2 = (PQ)^a = Q^2 = (QR)^b = ... = S^2 = (ST)^c = T^2,$$

$$\gamma^{-1}P\gamma = T.$$

For this the annotation is

$$^{\gamma}{*}^{P}\mathbf{a}^{Q}\mathbf{b}^{R}...^{S}\mathbf{c}^{T}.$$

Finally, a *crosscap* ($\times$) yields one Greek generator, δ, and one Latin generator, Z, subject to the relation

$$Z^2 = \delta$$

and annotated by

$$^{\delta}{\times}^{Z}.$$

This completes our description of the local relations. If there are h handles, g gyration points, k kaleidoscopes, and x crosscaps, then the only other relation is the global one:

$$\alpha_1\alpha_2...\alpha_h\beta_1\beta_2...\beta_g\gamma_1\gamma_2...\gamma_k\delta_1\delta_2...\delta_x = 1,$$

which we usually put first.

Understanding and Simplifying the Presentations: The Geometry of the Generators

The geometrical meaning of our generators is simpler to explain in any particular case than it is in general. So, we'll postpone the general description (and proof) to Chapter 14 and content ourselves for now with some illuminating examples.

Pure Kaleidoscope Groups

The generic presentation $^\gamma *^P 6^Q 3^R 2^S$ for $*\mathbf{632}$ is

$$\gamma = 1 = P^2 = (PQ)^6 = Q^2 = (QR)^3 = R^2 = (RS)^2 = S^2,$$

$$S = \gamma^{-1} P \gamma.$$

We can obviously simplify this by dropping γ and replacing S by P. This gives the simplified presentation $*^P 6^Q 3^R 2$:

$$1 = P^2 = (PQ)^6 = Q^2 = (QR)^3 = R^2 = (RP)^2,$$

in which the remaining generators, P, Q, and R, are just the reflections in the edges of the defining triangle. The relation $(PQ)^6 = 1$ is accounted for by the fact that the product PQ is a rotation through $2\pi/6$. Similarly, QR is a rotation by $2\pi/3$ and RP is a rotation by $2\pi/2$.

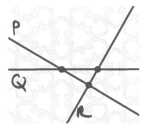

The groups

$$*\mathbf{632}, *\mathbf{442}, *\mathbf{333}, *\mathbf{2222}, *\mathbf{532}, *\mathbf{432}, *\mathbf{332}, *\mathbf{22N}, *\mathbf{NN}$$

all have this type of presentation.

Pure Gyration Groups

The three rotations $\alpha = PQ$, $\beta = QR$, and $\gamma = RP$ in $*\mathbf{632}$ generate its subgroup $\mathbf{632}$. They satisfy the presentation

$$^\alpha 6^\beta 3^\gamma 2 : \alpha\beta\gamma = 1 = \alpha^6 = \beta^3 = \gamma^2.$$

(For example, in the following figure α takes the red swoosh to the blue swoosh, then β takes this to the green swoosh, which γ sends back to our starting point, so that $\alpha\beta\gamma = 1$.)

The groups

$$\mathbf{632, 442, 333, 2222, 532, 432, 332, 22N, NN}$$

all have presentations of this form.

"Gyroscopic" Groups

The generic presentation for $\mathbf{4*2}$ is

$$^\alpha\mathbf{4}^\gamma *^P\mathbf{2}^Q : \alpha\gamma = 1 = \alpha^4 = P^2 = (PQ)^2 = Q^2, \gamma^{-1}P\gamma = Q.$$

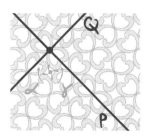

Here the generators α and γ are mutually inverse and are order-4 rotations that conjugate the reflections P and Q into each other:

$$Q = \gamma^{-1}P\gamma, P = \alpha^{-1}Q\alpha.$$

So, we can reduce the presentation to

$$^\alpha\mathbf{4} *^P\mathbf{2} : 1 = \alpha^4 = P^2 = (P\alpha P\alpha^{-1})^2$$

by dropping the generators γ and Q and the redundant relation $Q^2 = 1$.

The groups

$$4*2, 3*3, 3*2, 2*N$$

have similar presentations.

In **2*22** the kaleidoscopic part has two types of corner, exemplified by P, Q and Q, R. The generic presentation is

$$\alpha\gamma = 1 = P^2 = (PQ)^2 = Q^2 = (QR)^2 = R^2, \gamma^{-1}P\gamma = R,$$

which reduces to

$$1 = P^2 = (PQ)^2 = Q^2 = (Q\alpha P\alpha^{-1})^2.$$

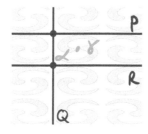

For **22*** there are two types of gyration. The generic presentation

$$^\alpha 2^\beta 2^\gamma *^P : \alpha\beta\gamma = 1 = P^2, P = \gamma^{-1}P\gamma$$

reduces to

$$^\alpha 2^\beta 2*^P : 1 = \alpha^2 = \beta^2 = P^2, \alpha\beta P = P\alpha\beta.$$

The combination $\gamma = \beta\alpha$ is a translation.

The group $\mathbf{N}*$ has a similar but simpler presentation:

$$^{\alpha}\mathbf{N}*^{P} : 1 = \alpha^{N} = P^{2}, \alpha P = P\alpha.$$

We have discussed all the plane and spherical groups except those that "involve topology":

$$**, *\times, \mathbf{22}\times, \times\times, \circ, \mathbf{N}\times.$$

The geometric meanings of the generators in these cases will be described as they arise in the next few chapters. However, we briefly discuss the first three here.

The generic presentation

$$^{\alpha}*^{P}\ ^{\beta}*^{Q} : \alpha\beta = 1 = P^{2} = Q^{2}, \alpha^{-1}P\alpha = P, \beta^{-1}Q\beta = Q$$

for $**$ simplifies to

$$^{\alpha}*^{P}*^{Q} : 1 = P^{2} = Q^{2}, \alpha P = P\alpha, \alpha Q = Q\alpha.$$

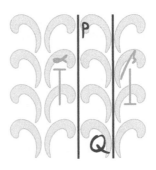

The presentation

$$^{\alpha}*^{P}\ ^{\beta}\times^{Z} : \alpha\beta = 1 = P^{2}, \alpha^{-1}P\alpha = P, \beta = Z^{2}$$

for $*\times$ reduces to

$$\alpha_{*}{}^{P}\times^{Z} : P^2 = 1, \alpha^{-1}P\alpha = P, \alpha^{-1} = Z^2,$$

and so to

$$_{*}{}^{P}\times^{Z} : 1 = P^2 = PZ^2PZ^{-2}.$$

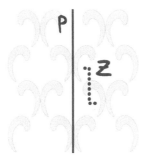

The presentation

$$2^{\alpha}2^{\beta}\;{}^{\gamma}\times^{Z} : \alpha\beta\gamma = \alpha^2 = \beta^2 = 1, \gamma = Z^2$$

for $22\times$ reduces to

$$2^{\alpha}2^{\beta}\times^{Z} : \alpha^2 = \beta^2 = \alpha\beta Z^2 = 1.$$

However, we will also find useful the "non-standard" presentation generated by Z and $Y = Z^{-1}\alpha$ with relations $(YZ)^2 = (YZ^{-1})^2 = 1$.

Where Are We?

We can now find a presentation in terms of generators and relations for the symmetry group of a pattern from its signature. Chapters 11–13 use this information to enumerate the symmetries of some families of colored patterns.

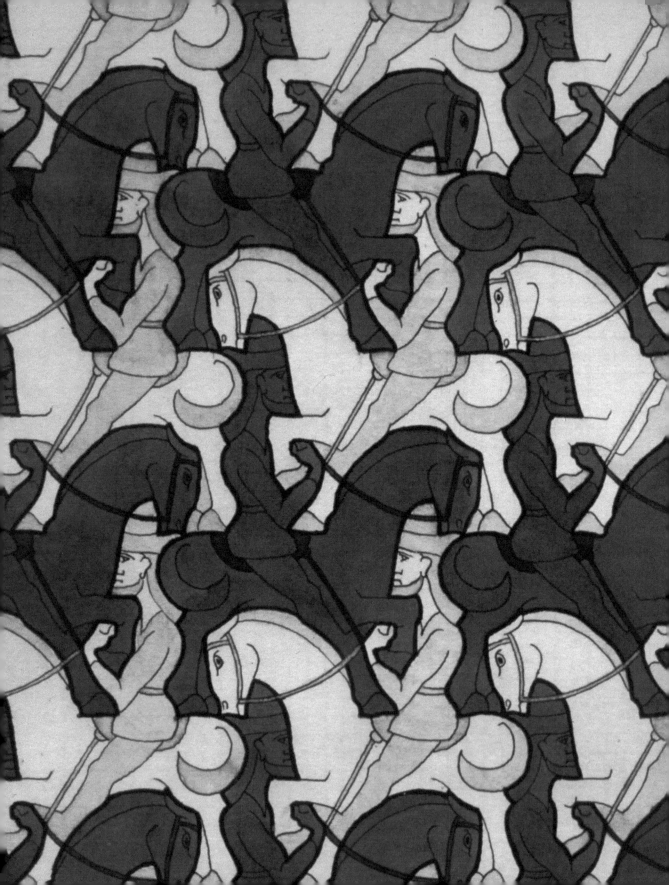

– 11 –

Twofold Colorations

Throughout the ages artists and artisans have made use of the wonderful ways color can interact with the symmetries of a pattern. In this chapter we discuss the mathematics of color symmetry. The situation is simplest in the two-color case, which we discuss first.

First, we'd better say exactly what we mean. M. C. Escher's *Symmetry Work 22* (at right) uses two colors but since the fish are of one color and the birds of another, this really makes no difference—a monochromatic version would have exactly the same symmetries. We are really studying patterns in which there are symmetries that interchange the two colors; we shall say that such a pattern has *twofold coloration*, and we will assign it a *(twofold) color signature*.

Symmetry Work 22, by M. C. Escher.

Describing Twofold Symmetries

In Escher's *Symmetry Work 67* (left) the horses and their riders are differently colored in a way that corresponds to the direction in which they are riding. What are the symmetries of this picture?

They are of two kinds. Half the symmetries fix each of the colors while the other half interchange them. The collection of all of the symmetries forms what we'll call the *full group* **G**, while the ones that fix the colors form the *kernel* **K**. We'll call it a **G/K** coloring and will often only indicate **G** and **K** by their signatures. So, for instance, *Symmetry Work 67* is a ××/○ coloring. This is because the figures of one color all face the same way, so its kernel consists only of translations (signature ○). But as the figure on the next page shows, the full group has signature ××, since there are also glide reflections that interchange the colors.

(opposite page) *Symmetry Work 67*, by M. C. Escher, is a twofold coloring.

More formally, we have the following definition. Two **G/K** and **G′/K′** colorings have the same *color type* just when **G** can be isotopically reshaped into **G′** in a way that takes **K** to **K′**. We usually regard colorings of the same type as identical. In many cases there is only one type of **G/K** coloring, so we can speak of "type **G/K**."

To completely describe twofold color types, we must say whether each symmetry fixes or interchanges the two colors a and b. Mathematically, this amounts to specifying a homomorphism from **G** into the group permuting {a, b} whose kernel is **K**, and we can do this very simply by replacing the generators in our presentation symbols by the appropriate permutations.

For example, the generators Y and Z in our simplified presentation $\times^Y \times^Z$ are glide reflections. Y and Z both interchange brown and yellow in the marginal figure, so a complete specification is

$$\times^{(\text{yellow} \leftrightarrow \text{brown})} \times^{(\text{yellow} \leftrightarrow \text{brown})};$$

we call this the *color signature* for the twofold coloring in *Symmetry Work 67*. Since the names of the two colors are immaterial, we lose no information by replacing these permutations by their orders,[1] which leads to a shorter form for the color signature:

$$\times^2 \times^2.$$

In the next few sections we shall classify all the types of twofold plane colorings. Although we must do this by considering all possibilities for their color signatures, it turns out that the simpler symbol **G/K** suffices to describe the color type in all but one case.

Glide reflections interchange the colors yellow and brown in *Symmetry Work 67*.

Classifying Twofold Plane Colorings

The symmetries of twofold colorings are completely determined by the ways in which the two colors are fixed or interchanged by the symmetries of the full group, or just by its generators. Each generator maps to a permutation of the colors, and they must do this in a manner so as to satisfy the relations obeyed by the generators. We warn the reader that the explanations sometimes get technical.

---

[1]The *order* of a permutation is the number of times it must be repeated for the colors to return to their original positions.

For example, we can find all possible twofold colorations of $*^P\mathbf{6}^Q\mathbf{3}^R\mathbf{2}$ by finding all mappings of P, Q, R to permutations p, q, r of the color-set {a,b} that satisfy the familiar relations $1 = p^2 = (pq)^6 = q^2 = (qr)^3 = r^2 = (rp)^2$. However, we can immediately narrow the possibilities to

$$*^2\mathbf{6}^2\mathbf{3}^2\mathbf{2}, \ *^1\mathbf{6}^2\mathbf{3}^2\mathbf{2}, \ *^2\mathbf{6}^1\mathbf{3}^1\mathbf{2}, \ *^1\mathbf{6}^1\mathbf{3}^1\mathbf{2},$$

because we can show that Q and R must in fact map to the same permutation, a fact that is indicated by "$Q = R$" in Table 11.1. This is because their images, q and r, lie in the group of order 2 generated by (y*ellow* ↔ b*rown*) and so satisfy $q^2 = r^2 = 1, qr = rq$, which together with $(qr)^3 = 1$ imply $q = r$. It's easy to check that for each of the possibilities mentioned above the images p, q, and r do indeed satisfy the relations, but since the last one has no symmetry interchanging the two colors, it is not "twofold". The other three cases are illustrated below, which shows that they are adequately described by

$$*\mathbf{632}/\mathbf{632}, \ *\mathbf{632}/\mathbf{3}*\mathbf{3}, \ *\mathbf{632}/*\mathbf{333}.$$

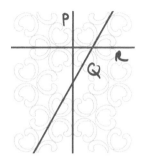

Generators for an (uncolored) pattern with symmetry type $*\mathbf{632}$

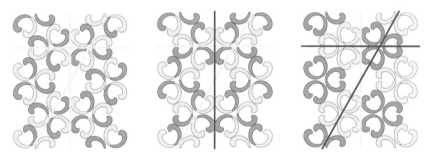

Later in this chapter, we will illustrate twofold coloring of all possible types. First, let's look at a few more examples and some special cases.

The symbol $^1\mathbf{2}^2\mathbf{2}*^1$ denotes a coloring of $^\alpha\mathbf{2}^\beta\mathbf{2}*^P$ in which α and P fix the colors, while β interchanges them. The figure below also gives the generators $(P, \alpha, \beta^{-1}\alpha\beta)$ for K in the resulting coloring, showing that the kernel for $^1\mathbf{2}^2\mathbf{2}*^1$ is also $\mathbf{22}*$, so it has type $\mathbf{22}*/\mathbf{22}*$.

The twofold coloring $^1\mathbf{2}^2\mathbf{2}*^1$ has type $\mathbf{22}*/\mathbf{22}*$.

The calculations for most cases are as simple as those above, but sometimes there are equivalences that are not immediately obvious. For example, **2222** has isotopic reshapings that achieve all permutations of α, β, γ, and δ, so that we get only one type of coloring from the six symbols

$$^1\mathbf{2}^1\mathbf{2}^2\mathbf{2}^2\mathbf{2}, \ ^1\mathbf{2}^2\mathbf{2}^1\mathbf{2}^2\mathbf{2}, \ ^1\mathbf{2}^2\mathbf{2}^2\mathbf{2}^1\mathbf{2}, \ ^2\mathbf{2}^1\mathbf{2}^1\mathbf{2}^2\mathbf{2}, \ ^2\mathbf{2}^1\mathbf{2}^2\mathbf{2}^1\mathbf{2}, \ ^2\mathbf{2}^2\mathbf{2}^1\mathbf{2}^1\mathbf{2},$$

explaining the "Number" entry, (6), in Table 11.1.

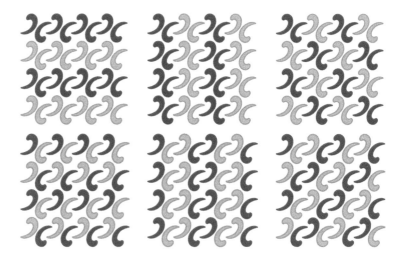

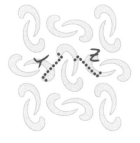

The equivalence of the two cases for $\mathbf{22}\times/\times\times$ is best shown using the presentation $\langle Y, Z \mid (YZ)^2 = (YZ^{-1})^2 = 1 \rangle$ on two orthogonal glide reflections. This shows that the two cases $Y \to +, Z \to -$ and $Y \to -, Z \to +$ are similar, although their names, $^2\mathbf{2}^2\mathbf{2}\times^1$ and $^2\mathbf{2}^2\mathbf{2}\times^2$, with respect to our usual presentation look different.

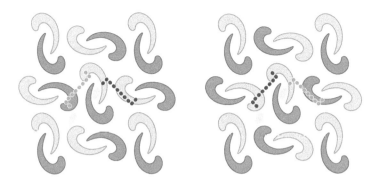

Finally, the equivalences for $\circ = \circ^{X,Y}$ are made visible by introducing the third translation Z defined by $XY = Z$, showing that all three cases

$$X \to +, Y \to -, Z \to -,$$

$$X \to -, Y \to +, Z \to -,$$

$$X \to -, Y \to -, Z \to +$$

are similar.

There is only one case in which the symbol $\mathbf{G/K}$ does not completely specify the color type. Namely, the case $**/**$ denotes two types that may be distinguished by their numbers of colorings.

Complete List of Twofold Color Types

Plane Patterns

The complete list of twofold color types for plane patterns appears in Table 11.1. For each of the 17 plane groups, we first give the annotated signature that specifies the presentation from which the color types are derived, followed by the short forms of our "color signatures." When more than one of these specify the same color

Annotated Signature and "Consequences"	Color Signature (short form)	Number (when > 1)	Color Type
	$*^1 6^2 3^2 2$		$*632/3*3$
$*^P 6^Q 3^R 2$	$*^2 6^1 3^1 2$		$*632/*333$
"$Q = R$"	$*^2 6^2 3^2 2$		$*632/632$
$^\alpha 6^\beta 3^\gamma 2$	$^2 6^1 3^2 2$		$632/333$
"$\beta = 1, \alpha = \gamma$"			
	$*^1 4^1 4^2 2,\ *^2 4^1 4^1 2$	(2)	$*442/*442$
	$*^1 4^2 4^2 2,\ *^2 4^2 4^1 2$	(2)	$*442/4*2$
$*^P 4^Q 4^R 2$	$*^1 4^2 4^1 2$		$*442/*2222$
	$*^2 4^1 4^2 2$		$*442/2*22$
	$*^2 4^2 4^2 2$		$*442/442$
	$^1 4 *^2 2$		$4*2/442$
$^\alpha 4 *^P 2$	$^2 4 *^1 2$		$4*2/2*22$
	$^2 4 *^2 2$		$4*2/22\times$
$^\alpha 4^\beta 4^\gamma 2$	$^1 4^2 4^2 2,\ ^2 4^1 4^2 2$	(2)	$442/442$
"$\alpha\beta\gamma = 1$"	$^2 4^2 4^1 2$		$442/2222$
$*^P 3^Q 3^R 3$	$*^2 3^2 3^2 3$		$*333/333$
"$P = Q = R$"			
$^\alpha 3 *^P 3$	$^1 3 *^2 3$		$3*3/333$
"$\alpha = 1$"			
$^\alpha 3^\beta 3^\gamma 3$		No case	
"$\alpha = \beta = \gamma = 1$"			
	$*^1 2^1 2^1 2^2 2,$ etc.	(4)	$*2222/*2222$
	$*^1 2^1 2^2 2^2 2,$ etc.	(4)	$*2222/2*22$
$*^P 2^Q 2^R 2^S 2$	$*^1 2^2 2^1 2^2 2,\ *^2 2^1 2^2 2^1 2$	(2)	$*2222/**$
	$*^1 2^2 2^2 2^2 2,$ etc.	(4)	$*2222/22*$
	$*^2 2^2 2^2 2^2 2$		$*2222/2222$
	$^1 2 *^1 2^2 2,\ ^1 2 *^2 2^1 2$	(2)	$2*22/22*$
	$^1 2 *^2 2^2 2$		$2*22/2222$
$^\alpha 2 *^P 2^Q 2$	$^2 2 *^1 2^1 2$		$2*22/*2222$
	$^2 2 *^1 2^2 2,\ ^2 2 *^2 2^1 2$	(2)	$2*22/*\times$
	$^2 2 *^2 2^2 2$		$2*22/22\times$
	$^1 2^1 2 *^2$		$22*/2222$
	$^1 2^2 2 *^1,\ ^2 2^1 2 *^1$	(2)	$22*/22*$
$^\alpha 2^\beta 2 *^P$	$^1 2^2 2 *^2,\ ^2 2^1 2 *^2$	(2)	$22*/22\times$
	$^2 2^2 2 *^1$		$22*/**$
	$^2 2^2 2 *^2$		$22*/\times\times$

Table 11.1. Table of twofold color types for plane patterns.

type, they are put on one line, followed by their number, which is the number of colorings of this type for the given group. The diagrams on pages 142–143 display examples of all the twofold color types of plane patterns.

Annotated Signature and "Consequences"	Color Signature (short form)	Number (when > 1)	Color Type
$\alpha\mathbf{2}^\beta\mathbf{2}\times^Z$	$^1\mathbf{2}^1\mathbf{2}\times^2$		**22×/2222**
	$^2\mathbf{2}^2\mathbf{2}\times^1,\ ^2\mathbf{2}^2\mathbf{2}\times^2$	(2)	**22×/××**
$\alpha\mathbf{2}^\beta\mathbf{2}^\gamma\mathbf{2}^\delta\mathbf{2}$	$^1\mathbf{2}^1\mathbf{2}^2\mathbf{2}^2\mathbf{2}$, etc.	(6)	**2222/2222**
	$^2\mathbf{2}^2\mathbf{2}^2\mathbf{2}^2\mathbf{2}$		**2222/∘**
	$^1{}_*2{}_*2$		∗∗/∘
	$^1{}_*1{}_*2,\ ^1{}_*2{}_*1$	(2)	∗∗/∗∗(2)
$\alpha{}_*P{}_*Q$	$^2{}_*1{}_*1$	(1)	∗∗/∗∗(1)
	$^2{}_*1{}_*2,\ ^2{}_*2{}_*1$	(2)	∗∗/∗×
	$^2{}_*2{}_*2$		∗∗/××
	${}_*1\times^2$		∗×/∗∗
$\alpha{}_*P\times^Z$	${}_*2\times^1$		∗×/××
	${}_*2\times^2$		∗×/∘
$\times^Y\times^Z$	$\times^1\times^2,\ \times^2\times^1$	(2)	××/××
	$\times^2\times^2$		××/∘
$\circ^{X,Y}$	$\circ^{1,2},\ \circ^{2,1},\ \circ^{2,2}$	(3)	∘/∘

Table 11.1. (continued.)

The final column of the table gives the "**G/K**" notation, which specifies the color type except in the case ∗∗/∗∗, which we now describe.

The ∗∗/∗∗ coloring below is unique (up to swapping the two colors), so we call its type ∗∗/∗∗ (1).

However, the two ∗∗/∗∗ colorings that follow have the same type, which we therefore call ∗∗/∗∗ (2).

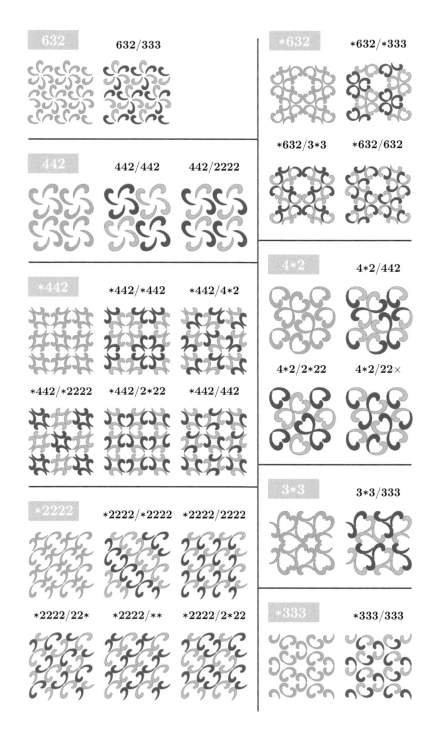

2*22	2*22/*2222	2222	22×	××

2*22/22*	2*22/2222	2222/○	22×/2222	××/○

2*22/*×	2*22/22×	2222/2222	22×/××	××/××

22*	22*/××	22*/**	*×	*×/**

22*/22×	22*/2222	22*/22*	*×/○	*×/××

**	**/**(1)	**/**(2)	○	○/○

**/*×	**/○	**/××		

The Topology of Coloring Symmetries

If H is a color subgroup of G, then the orbifold of G must be a "branched cover" of the orbifold of H. That is, we can wrap H's orbifold around that of G.

For example, the full type *2222 can be two-colored to have color type *2222/2222. The orbifold of 2222, a sphere with four cone points, can be squashed flat, double covering the orbifold for *2222: just as the group is half as large, the orbifold is twice as big.

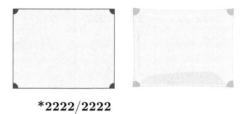

*2222/2222

For simple color types, at least, we can easily intuit such covers. For example, each of these orbifolds can be double-covered by the orbifold for **2222**, and each of the corresponding symmetry groups contains an index-2 group of type **2222**.

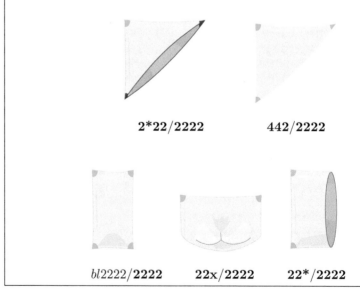

2*22/2222   **442/2222**

*bl*2222/**2222**   **22x/2222**   **22*/2222**

Spherical and Frieze Patterns

Twofold color types for the sphere are computed similarly to those for the plane. The results of these computations are presented in Table 11.2, and the color types are illustrated on the following two pages. Each of the infinite families of spherical color types naturally corresponds to the color type of a frieze pattern. The seven twofold frieze types are just the cases of Table 11.2 with $N = M = \infty$. They are displayed on page 148.

Group	Color Signature	Number	Type
$*\mathbf{532}$	$*^2\mathbf{5}^2\mathbf{3}^2\mathbf{2}$		$*\mathbf{532/532}$
$\mathbf{532}$		no case	
$*\mathbf{432}$	$*^2\mathbf{4}^2\mathbf{3}^2\mathbf{2}$		$*\mathbf{432/432}$
	$*^1\mathbf{4}^2\mathbf{3}^2\mathbf{2}$		$*\mathbf{432/3}*\mathbf{2}$
	$*^2\mathbf{4}^1\mathbf{3}^1\mathbf{2}$		$*\mathbf{432}/*\mathbf{332}$
$\mathbf{432}$	$^2\mathbf{4}^1\mathbf{3}^2\mathbf{2}$		$\mathbf{432/332}$
$*\mathbf{332}$	$*^2\mathbf{3}^2\mathbf{3}^2\mathbf{2}$		$*\mathbf{332/332}$
$\mathbf{3}*\mathbf{2}$	$^1\mathbf{3}*^2\mathbf{2}$		$\mathbf{3}*\mathbf{2/332}$
$\mathbf{332}$		no case	
$*\mathbf{22N}$	$*^1\mathbf{2}^2\mathbf{2}^1\mathbf{N}$		$*\mathbf{22N}/*\mathbf{NN}$
	$*^2\mathbf{2}^1\mathbf{2}^2\mathbf{N}$		$*\mathbf{22N/N}*$
	$*^2\mathbf{2}^2\mathbf{2}^2\mathbf{N}$		$*\mathbf{22N/22N}$
	$*^2\mathbf{2}^1\mathbf{2}^1\mathbf{N} = *^1\mathbf{2}^1\mathbf{2}^2\mathbf{N}$	(2)	$*\mathbf{22N}/*\mathbf{22M}$
	$*^2\mathbf{2}^2\mathbf{2}^1\mathbf{N} = *^1\mathbf{2}^2\mathbf{2}^2\mathbf{N}$	(2)	$*\mathbf{22N/2}*\mathbf{M}$
$\mathbf{2}*\mathbf{N}$	$^2\mathbf{2}*^1\mathbf{N}$		$\mathbf{2}*\mathbf{N}/*\mathbf{NN}$
	$^1\mathbf{2}*^2\mathbf{N}$		$\mathbf{2}*\mathbf{N/22N}$
	$^2\mathbf{2}*^2\mathbf{N}$		$\mathbf{2}*\mathbf{N/N}\times$
$\mathbf{22N}$	$^2\mathbf{2}^2\mathbf{2}^1\mathbf{N}$		$\mathbf{22N/NN}$
	$^1\mathbf{2}^2\mathbf{2}^2\mathbf{N} =^2 \mathbf{2}^1\mathbf{2}^2\mathbf{N}$	(2)	$\mathbf{22N/22M}$
$*\mathbf{NN}$	$*^2\mathbf{N}^2\mathbf{N}$		$*\mathbf{NN/NN}$
	$*^2\mathbf{N}^1\mathbf{N} = *^1\mathbf{N}^2\mathbf{N}$	(2)	$*\mathbf{NN}/*\mathbf{MM}$
$\mathbf{N}*$	$^1\mathbf{N}*^2$		$\mathbf{N}*\mathbf{/NN}$
	$^2\mathbf{N}*^1$		$\mathbf{N}*\mathbf{/M}*$
	$^2\mathbf{N}*^2$		$\mathbf{N}*\mathbf{/M}\times$
$\mathbf{N}\times$	$^1\mathbf{N}\times^2$		$\mathbf{N}\times\mathbf{/NN}$
$\mathbf{NN}$	$^2\mathbf{N}^2\mathbf{N}$		$\mathbf{NN/MM}$

Table 11.2. Twofold spherical types (here $M = N/2$).

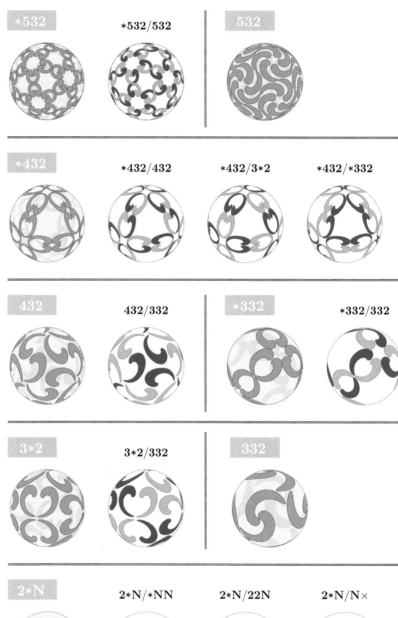

*532   *532/532   532

*432   *432/432   *432/3*2   *432/*332

432   432/332   *332   *332/332

3*2   3*2/332   332

2*N   2*N/*NN   2*N/22N   2*N/N×

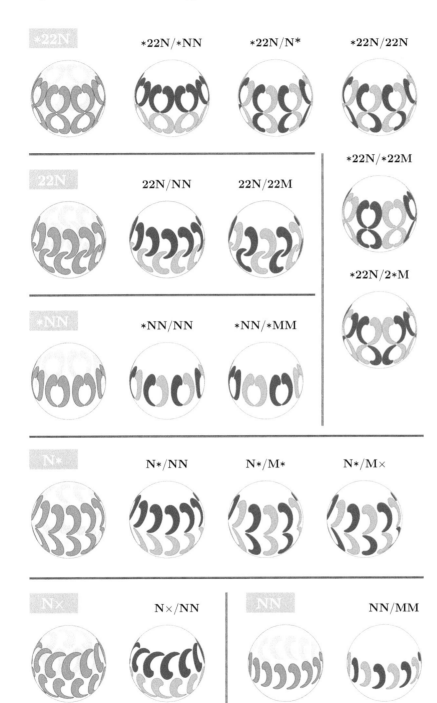

2*∞   2*∞/*∞∞   2*∞/22∞   2*∞/∞×

*22∞   *22∞/*∞∞   *22∞/∞*   *22∞/22∞

*22∞/*22∞

22∞   22∞/∞∞   22∞/22∞

22∞/2∞

*∞∞   *∞∞/∞∞   *∞∞/*∞∞

∞*   ∞*/∞∞   ∞*/∞*   ∞*/∞×

∞×   ∞×/∞∞

∞∞   ∞∞/∞∞

Duality Groups

When a polyhedron or tessellation is dual to a copy of itself, we obtain a more symmetrical figure by drawing both of the corresponding maps on the same surface in two different colors, say green and brown. If the resulting figure has color type $\mathbf{G/H}$, we'll say that the *duality group* is $\mathbf{H\backslash G}$. We reverse the order since now $\mathbf{H}$ is the more important group and also to distinguish between duality groups and more general color-symmetry groups.

Not every twofold color group $\mathbf{G/H}$ yields a duality group $\mathbf{H\backslash G}$. The reason is that a point P fixed by an element of order 3 or more in $\mathbf{H}$ cannot also be fixed by any element of $\mathbf{G}$ outside $\mathbf{H}$. This is because P must be either a vertex or the center of a face, but the elements of $\mathbf{G}$ outside $\mathbf{H}$ (the "dualities") interchange those concepts.

For instance, the green and brown pyramids in Figure 11 are mutually dual. In this case all the symmetries of one fix its apex, but the dualities interchange it with the apex of the other (and vice versa).

This condition prevents $\mathbf{432\backslash *432}$ from being a duality group since the fourfold gyration points of $\mathbf{H} = \mathbf{432}$ are fixed by further elements of $*\mathbf{432}$. The groups that *do* satisfy it *can* all arise as duality groups and are listed in Table 11.3.

A Euclidean tessellation that is dual to a copy of itself has a duality group $\mathbf{H\backslash G}$ in the same way as the spherical case. The Euclidean duality groups are listed in Table 11.4.

Figure 11.1. Two intersecting pyramids.

$$*332 \setminus *432$$
$$332 \setminus 432$$
$$332 \setminus 3*2$$

$$*222 \text{ or } 2*2 \setminus * 224$$
$$222 \setminus 224$$
$$*22 \setminus *44$$
$$2* \text{ or } 2\times \setminus 4*$$
$$22 \setminus 44$$

$$*22 \text{ or } 2* \text{ or } 222 \text{ or } *22 \setminus *222$$
$$*22 \text{ or } 222 \text{ or } 2\times \setminus 2*2$$
$$22 \setminus 222$$
$$22 \text{ or } * \setminus *22$$
$$22 \text{ or } * \text{ or } \times \setminus 2*$$
$$22 \setminus 2\times$$
$$1 \setminus 22$$
$$1 \setminus *$$
$$1 \setminus \times$$

Table 11.3. The spherical duality groups.

$$*442 \text{ or } *2222 \text{ or } 2*22 \setminus *442$$
$$442 \text{ or } 2*22 \text{ or } 22\times \setminus 4*2$$
$$442 \text{ or } 2222 \setminus 442$$

$$*2222 \text{ or } 2*22 \text{ or } ** \text{ or } 22* \text{ or } 2222 \setminus *2222$$
$$22* \text{ or } 2222 \text{ or } *2222 \text{ or } *\times \text{ or } 22\times \setminus 2*22$$
$$2222 \text{ or } 22* \text{ or } 22\times \text{ or } ** \text{ or } \times\times \setminus 22*$$
$$2222 \text{ or } \times\times \setminus 22\times$$
$$2222 \text{ or } \bigcirc \setminus 2222$$
$$\bigcirc \text{ or } **(1) \text{ or } **(2) \text{ or } *\times \text{ or } \times\times \setminus **$$
$$** \text{ or } \times\times \text{ or } \bigcirc \setminus *\times$$
$$\times\times \text{ or } \bigcirc \setminus \times\times$$
$$\bigcirc \setminus \bigcirc$$

Table 11.4. The Euclidean duality groups.

Where Are We?

In the previous chapter we described the algebraic structure underlying the symmetric patterns that we've been studying. In this chapter we defined twofold colorings and used our algebraic understanding of repeating patterns to enumerate twofold color types for plane, spherical, and frieze patterns. Using that enumeration, we also enumerated the duality groups that arise when a polyhedron or tessellation can be dual to a copy of itself. Subsequent chapters deal with threefold and primefold colorings.

– 12 –

Threefold Colorings of Plane Patterns

What *is* a "color"? Although "color-symmetry" is the standard term for this subject, the number of colors a pattern has is a rather indefinite concept. The problem arises particularly when there are three or more colors, although really it's already happened in the two color case.

Let's look again at Escher's *Symmetry Work 67* of horses and their riders. In Chapter 11, we described this as a pattern that has two colors, brown and yellow. But it has several shades of brown and several of yellow, and the exact number of colors is rather indefinite— we might even think it infinite if we regard the colors as continuously variable. So, it isn't really a "two-color pattern," which is why we deliberately used the term "twofold color symmetry" where other authors might say "two-color symmetry."

However, you know what they mean by calling it a two-color pattern: there is a clear distinction between "brownish" regions and "yellowish" regions even though each involves several distinct shades. Lots of patterns (for example, Escher's *Symmetry Work 72* with sailboats and rainbow-colored fish) have many colors, but their symmetries never change the colors. So, from our point of view, the colors are irrelevant, and they are "onefold" colorings. In general, an n-fold coloring is one in which there are regions of n distinct "colors" (which need not in fact be single colors) that has symmetries that take any one of the n "colors" to any other one. (Mathematicians express this by saying that the symmetries are transitive on the colors.) These "colors" need not cover the whole pattern; for instance,

Symmetry Work 67, by M. C. Escher.

(opposite page) Escher's *Symmetry Work 70* is a threefold coloring of type $632^3/632/333$, which we abbreviate **632//333**.

they might be separated by black lines or white space (to be regarded as "uncolored"), and we can even allow the "colors" to overlap.

A Look at Threefold Colorings

Having found all the twofold coloring types in the plane, we might move on to describe the threefold ones. It is only slightly more difficult to describe the p-fold types for arbitrary primes p, so in the next chapter we will do that as well. We work with three arbitrary colors A, B, and C—say Apple green, Banana yellow, and Cherry red.

In this case, each symmetry achieves a permutation of the three colors, as in Figure 12.1. Therefore, the coloring type is determined by a homomorphism from $\mathbf{G}$ onto some subgroup of the group

$$\mathbf{S}[3] = \{1, (A)(BC), (B)(AC), (C)(AB), (ABC), (CBA)\}$$

of all permutations of the three colors.

Moreover, since our term "threefold coloring" applies only to the "transitive" cases when there are symmetries that take any one color to any other color, the subgroup can only be $\mathbf{S}[3]$ itself or its cyclic subgroup

$$\mathbf{C}[3] = \{1, (ABC), (CBA)\}$$

(because these are the only transitive subgroups of $\mathbf{S}[3]$).

In Chapter 11 there were two groups: the full group $\mathbf{G}$ of all symmetries and the kernel $\mathbf{K}$ containing just the symmetries that fix each

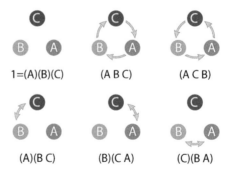

Figure 12.1. Threefold coloring types.

color. Now there is a third group **H**, consisting of the symmetries that fix any one chosen color. This group is called the *stabilizer* of that color. (Actually, there are three stabilizer groups, one for each color, but they are all abstractly equivalent—indeed *conjugate* in the group-theoretical sense.)

The right-hand figure above is a coloring of a pattern of type ∗**333**. On the left, we ignore the three colors: the group **G** has signature ∗**333**. In the middle, we consider the group **H** of symmetries that fix color C: it has signature **3∗3**. Finally, at right, the signature of the group **K** of symmetries that fix all three colors is **333**.

The signatures of these groups usually specify the color type, which we therefore call **G**^{3}/**H**/**K**. We simplify this to **G**^{3}/**K** when **H** = **K** and to **G**^{3}//**K** otherwise, since we don't usually need to specify **H**. For p-fold colorings, we use similar notations, **G**^{p}/**H**/**K**, **G**^{p}/**K**, and **G**^{p}//**K**, and allow ourselves to omit the number p when it is understood.

Complete List for Plane Patterns

Table 12.1 lists all of the threefold colorings for plane patterns. We use 1 for $(A)(B)(C)$, the identity permutation; 2 for any of $(A)(BC)$, $(B)(CA)$, or $(C)(AB)$, the permutations of order 2; and 3 for (ABC) or (CBA), those of order 3.

In the following we evaluate these groups **G**, **H**, **K** in the various cases and only telegraphically hint at the arguments (often quite technical) that restrict us to the answers given.

Annotated Signature	Color Signature	Type
$*^P 6^Q 3^R 2$	$*^1 6^2 3^2 2$	$*\mathbf{632}//*\mathbf{333}$
	$*^2 6^2 3^2 2$	$*\mathbf{632}//\mathbf{2222}$
$^\alpha 6^\beta 3^\gamma 2$	$^3 6^2 3^2 2$	$\mathbf{632}/\mathbf{2222}$
		$\mathbf{632}//\mathbf{333}$
$*^P 3^Q 3^R 3$	$*^{(AB)} 3^{(BC)} 3^{(CA)} 3$	$*\mathbf{333}//\bigcirc$
	$*^{(AB)} 3^{(BC)} 3^{(BC)} 3$	$*\mathbf{333}//\mathbf{333}$
$^\alpha 3 *^P 3$	$^3 3 *^1 3$	$\mathbf{3}*\mathbf{3}/\bigcirc$
	$^3 3 *^2 3$	$\mathbf{3}*\mathbf{3}//*\mathbf{333}$
$^\alpha 3^\beta 3^\gamma 3$	$^3 3^3 3^3 3$	$\mathbf{333}/\bigcirc$
	$^3 3^3 3^1 3$	$\mathbf{333}/\mathbf{333}$
$*^P 2^Q 2^R 2^S 2$	$*^1 2^2 2^1 2^2 2$	$*\mathbf{2222}//*\mathbf{*}$
$^\alpha 2 *^P 2^Q 2$	$^2 2 *^1 2^2 2$	$\mathbf{2}*\mathbf{22}//*\mathbf{*}$
$^\alpha 2^\beta 2 *^P$	$^1 2^1 2 *^2$	$\mathbf{22}*//\bigcirc$
	$^2 2^2 2 *^1$	$\mathbf{22}*//*\mathbf{*}$
$^\alpha 2^\beta 2 \times^Z$	$^2 2^2 2 \times^2$	$\mathbf{22}\times//\times\times$
$^\alpha 2^\beta 2^\gamma 2^\delta 2$	$^2 2^2 2^2 2^2 2$	$\mathbf{2222}//\bigcirc$
$^\alpha *^P *^Q$	$^3 *^1 *^1$	$*\mathbf{*}/*\mathbf{*}$
	$^1 *^2 *^2$	$*\mathbf{*}//\bigcirc$
$^\alpha *^P \times^Z$	$*^1 \times^3$	$*\times/*\times$
	$*^2 \times^2$	$*\times//\bigcirc$
$\times^Y \times^Z$	$\times^3 \times^3$	$\times\times/\times\times$
	$\times^2 \times^2$	$\times\times//\bigcirc$
$\bigcirc^{X,Y}$	$\bigcirc^{1,3}$	$\bigcirc/\bigcirc$

Table 12.1. Threefold coloring types on the plane.

$*^P 6^Q 3^R 2 : 1 = P^2 = (PQ)^6 = Q^2 = (QR)^3 = R^2 = (RP)^2.$

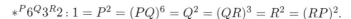

Since $P, Q, R \to 1$ or 2, there is no map onto $\mathbf{C}[3]$. To generate $\mathbf{S}[3]$, two of them must map to distinct 2's, but these can't be P and R in view of $(RP)^2 = 1$. The cases in which $Q, R \to 1, 2$ or $2, 1$ violate $(QR)^3 = 1$, so only $*^1 6^{(AB)} 3^{(BC)} 2 = *^1 6^2 3^2 2$, type $*\mathbf{632}//*\mathbf{333}$, and $*^{(BC)} 6^{(AB)} 3^{(BC)} 2 = *^2 6^2 3^2 2$, type $*\mathbf{632}//\mathbf{2222}$, survive.

$^\alpha 6^\beta 3^\gamma 2 : \alpha\beta\gamma = 1 = \alpha^6 = \beta^3 = \gamma^2.$

If there is a map onto $\mathbf{C}[3]$ then $\gamma \to 1$, so $\alpha\beta \to 1$. Thus, without loss of generality, $\alpha \to (ABC)$ $\beta \to (CBA)$, giving $^{(ABC)}\mathbf{6}^{(CBA)}\mathbf{3}^1\mathbf{2} = {}^3\mathbf{6}^3\mathbf{3}^1\mathbf{2}$, type $\mathbf{632/2222}$. To generate $\mathbf{S}[3]$, α or $\gamma \to 2$, since $\beta \to 1$ or 3. Now $\alpha\beta\gamma = 1$ determines β from α and γ, which must therefore map to distinct 2's, yielding one more case: $^{(AB)}\mathbf{6}^{(ABC)}\mathbf{3}^{(AC)}\mathbf{2} = {}^2\mathbf{6}^3\mathbf{3}^2\mathbf{2}$, type $\mathbf{632//333}$.

$*^P4^Q4^R2 : 1 = P^2 = (PQ)^4 = Q^2 = (QR)^4 = R^2 = (RP)^2.$ No cases! Since $P, Q, R \to 1$ or 2, none map onto $\mathbf{C}[3]$. To generate $\mathbf{S}[3]$, one needs two distinct 2's; but their product maps to 3, contradicting a relation.

$^\alpha 4*^P2 : 1 = \alpha^4 = P^2 = (P\alpha P\alpha^{-1})^2.$ No cases! As before, none map onto $\mathbf{C}[3]$. None map onto $\mathbf{S}[3]$ since the only hope is $\alpha \to (AB), P \to (BC)$, which contradicts $(P\alpha P\alpha^{-1})^3 = 1$.

$^\alpha 4^\beta 4^\gamma 2 : \alpha\beta\gamma = 1 = \alpha^4 = \beta^4 = \gamma^2.$ No cases! None onto $\mathbf{C}[3]$ and none onto $\mathbf{S}[3]$ since two of $\alpha, \beta, \gamma \to$ distinct 2's, making their product (which is the inverse of the third 2) $\to 3$.

$*^P3^Q3^R3 : 1 = P^2 = (PQ)^3 = Q^2 = (QR)^3 = R^2 = (RP)^3.$

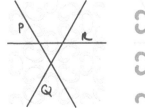

The fact that $P, Q, R \to 1$ or 2 implies that none map onto $\mathbf{C}[3]$. Now, without loss of generality, $P, Q \to$ distinct 2's, when $(QR)^3 = 1$ implies $R \not\to 1$, yielding (without loss of generality) two cases: $*^{(AB)}\mathbf{3}^{(BC)}\mathbf{3}^{(CA)}\mathbf{3}$, type $*\mathbf{333}//\circ$, and $*^{(AB)}\mathbf{3}^{(BC)}\mathbf{3}^{(BC)}\mathbf{3}$, type $*\mathbf{333}//\mathbf{333}$. (We can't use the short notation here, since both of these abbreviate to $*^2\mathbf{3}^2\mathbf{3}^2\mathbf{3}$.)

$$^{\alpha}3*^{P}3 : 1 = \alpha^3 = P^2 = (P\alpha P\alpha^{-1})^3.$$

Since $\alpha \to 1$ or 3, $p \to 1$ or 2, and there are just two cases: onto $\mathbf{C}[3]$ $^{(ABC)}\mathbf{3}*^1\mathbf{3} = {}^3\mathbf{3}*^1\mathbf{3}$, type $\mathbf{3}*\mathbf{3}/\circ$, and onto $\mathbf{S}[3]$ $^{(ABC)}\mathbf{3}*^{(AB)}\mathbf{3} = {}^3\mathbf{3}*^2\mathbf{3}$, type $\mathbf{3}*\mathbf{3}//\mathbf{333}$.

$$^{\alpha}3^{\beta}3^{\gamma}3 : \alpha\beta\gamma = 1 = \alpha^3 = \beta^3 = \gamma^3.$$

Since $\alpha, \beta, \gamma \to 1$ or 3, the image must be $\mathbf{C}[3]$. Now $\alpha\beta\gamma = 1$ implies just two cases (w.l.o.g.): $^{(ABC)}\mathbf{3}^{(ABC)}\mathbf{3}^{(ABC)}\mathbf{3} = {}^3\mathbf{3}^3\mathbf{3}^3\mathbf{3}$, type $\mathbf{333}/\circ$, and $^{(ABC)}\mathbf{3}^{(CBA)}\mathbf{3}^1\mathbf{3} = {}^3\mathbf{3}^3\mathbf{3}^1\mathbf{3}$, type $\mathbf{333}/\mathbf{333}$.

Where Are We?

We have enumerated the threefold color types of plane groups except for

$$*\mathbf{2222}, \mathbf{2}*\mathbf{22}, \mathbf{22}*, \mathbf{22}\times, \mathbf{2222}, **, *\times, \times\times, \circ.$$

We could continue by handling those in the same manner, but the more sophisticated argument of the next chapter finds the p-fold types for larger primes at the same time. However, for the reader's convenience, we have quoted the results in Table 12.1.

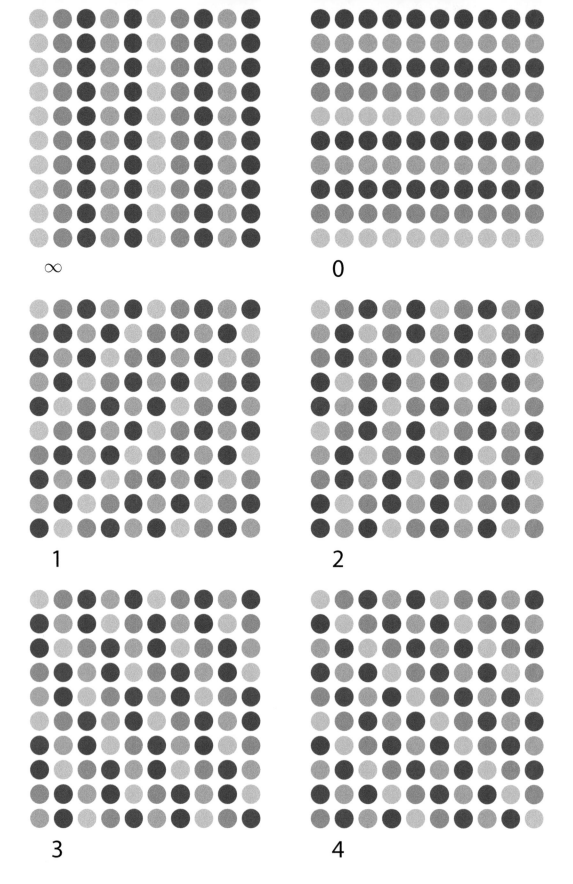

∞

0

1

2

3

4

– 13 –

Other Primefold Colorings

Plane Patterns

We next show how to enumerate the p-fold colorings of plane patterns for any prime p. The new argument uses a fair amount of group theory. In group theorists' slang, the group $\mathbf{G}$ of a Euclidean plane pattern has shape $\mathbf{L.Q}$, meaning that it has a normal subgroup $\mathbf{L}$ with finite quotient $\mathbf{Q}$. The group $\mathbf{L} \cong \mathbf{C}_\infty \times \mathbf{C}_\infty$ ("the lattice") consists of the translations in $\mathbf{G}$, and its quotient group $\mathbf{Q}$ ("the point group") is not necessarily realized as a subgroup. The argument works for a prime p that doesn't divide the order of $\mathbf{Q}$ (so, for $p \geq 5$ in the cases $*\mathbf{632}$, $\mathbf{632}$, $*\mathbf{333}$, $\mathbf{3}*\mathbf{3}$, and $\mathbf{333}$, and for $p \geq 3$ otherwise.) The missing threefold cases are precisely those that were done in the previous chapter.

We need only seek the stabilizer subgroup $\mathbf{H}$ of index p in $\mathbf{G}$ that fixes a given color, since this determines the coloring type. Since $\mathbf{H}$ has index p in $\mathbf{G}$, it must contain the normal subgroup $\mathbf{L}^p$ of $\mathbf{G}$ that is generated by the pth powers of all translations. $\mathbf{H}$ is therefore

Figure 13.1. In the images in this chapter, we will use $p = 5$, with this coloring. Note that $\langle -t \rangle$ is the permutation $(0)(14)(23)$ and $\langle 1 - t \rangle$ is $(01)(24)(3)$.

(opposite page) There are just $p + 1$ groups of index p in $\mathbf{L}$ (see page 162), determined by their slopes modulo p. Here, $p = 5$, and we show groups with slopes ∞, 0, 1, 2, 3, and 4.

determined by its image modulo $\mathbf{L}^p$, a subgroup of index p in the finite group $\mathbf{G}/\mathbf{L}^p \cong (\mathbf{C}_p \times \mathbf{C}_p) \cdot \mathbf{Q}$.

But, we can now regard $\mathbf{Q}$ as a particular subgroup[1] of this finite group; this must be in $\mathbf{H}$, and so $\mathbf{H}$ is determined by its intersection with $\mathbf{L}$, a subgroup of index p in $\mathbf{L}$.

There are just $p+1$ such subgroups of $\mathbf{L}$, characterized by their "slopes" (see page 283), but ours must be one that is fixed by $\mathbf{Q}$. The groups

$$*\mathbf{2222}, \mathbf{2}*\mathbf{22}, \mathbf{22}*, \mathbf{22}\times, **, *\times, \times\times$$

have symmetries that negate slopes, which restricts us to the two possibilities[2] corresponding to slopes ∞ and 0. We telegraph the arguments in the following, as before; the typical "color" t will be one of $0, 1, 2, ..., p-1$, and the permutation that takes t to $f(t)$ is called $\langle f(t) \rangle$.

$*^P\mathbf{2}^Q\mathbf{2}^R\mathbf{2}^S\mathbf{2}$ $1 = P^2 = (PQ)^2 = Q^2 = (QR)^2 = R^2 = (RS)^2 = S^2 = (SP)^2$. Here, the only invariant slopes ∞ and 0 are interchangeable, so yield just one case: $P \to 1$, $Q \to \langle 1-t \rangle$, $R \to 1$, $S \to \langle -t \rangle$, i.e., $*^1\mathbf{2}^2\mathbf{2}^1\mathbf{2}^2\mathbf{2}$, type $*\mathbf{2222}//**$.

$^\alpha\mathbf{2}*^P\mathbf{2}^Q\mathbf{2}$ $1 = \alpha^2 = P^2 = (PQ)^2 = Q^2 = (QR)^2 = \alpha^{-1}P\alpha R^{-1}$. The same argument yields one case: $\alpha \to \langle -t \rangle$, $P \to 1$, $Q \to \langle 1-t \rangle$, i.e., $^2\mathbf{2}*^1\mathbf{2}^2\mathbf{2}$, type $\mathbf{2}*\mathbf{22}//**$.

$^\alpha\mathbf{2}^\beta\mathbf{2}*^P$ $\alpha\beta\gamma = 1 = \alpha^2 = \beta^2 = P^2 = \gamma^{-1}P\gamma P^{-1}$ Here, the two invariant slopes ∞ and 0 yield two possibilities that lead to distinct cases: $\alpha \to 1, \beta \to 1, P \to \langle 1-t \rangle$, i.e., $^1\mathbf{2}^1\mathbf{2}*^2$ of type $\mathbf{22}*//\circ$, and $\alpha \to \langle -t \rangle, \beta \to \langle 1-t \rangle$, $P \to 1$, or $^2\mathbf{2}^2\mathbf{2}*^1$, type $\mathbf{22}*//**$.

---

[1]Because, by Hall's theorem, there *is* a subgroup of the same order as $\mathbf{Q}$, and all such are conjugate. (Hall's theorem asserts that, for any set π of primes, a solvable finite group $\mathbf{F}$ has a subgroup whose order is the π part of $|\mathbf{F}|$ and that all such subgroups are conjugate.)

[2]These may, however, be equivalent under some automorphism that interchanges x and y.

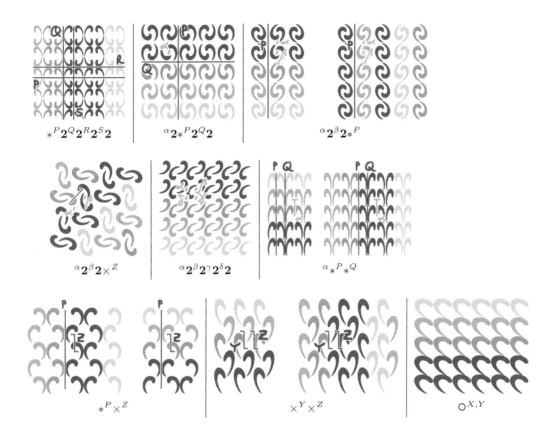

$^\alpha\mathbf{2}^\beta\mathbf{2}\times^Z$ — $\alpha\beta\gamma = 1 = \alpha^2 = \beta^2 = Z^2\gamma^{-1}$. The slopes ∞ and 0 are interchangeable, so we get just one case: $\alpha, \beta \to \langle -t \rangle$, $Z \to \langle 1 - t \rangle$, or $^2\mathbf{2}^2\mathbf{2}\times^2$, type $\mathbf{22}\times//\times\times$.

$^\alpha\mathbf{2}^\beta\mathbf{2}^\gamma\mathbf{2}^\delta\mathbf{2}$ — $\alpha\beta\gamma\delta = 1 = \alpha^2 = \beta^2 = \gamma^2 = \delta^2$. Here, the elements of the point group either fix or negate all vectors. Therefore, all $p+1$ slopes are invariant, but the group has automorphisms that make them all equivalent. We obtain just one case: $\alpha, \beta \to \langle -t \rangle$, $\gamma, \delta \to \langle 1 - t \rangle >$, or $^2\mathbf{2}^2\mathbf{2}^2\mathbf{2}^2\mathbf{2}$, type $\mathbf{2222}//\circ$.

$^\alpha\!\ast^P\!\ast^Q$ — $1 = P^2 = Q^2 = \alpha^{-1}P\alpha P^{-1} = \alpha Q\alpha^{-1}Q^{-1}$. The slopes ∞ and 0 lead to distinct cases: $\alpha \to \langle t + 1 \rangle$, $P, Q \to 1$, or $^p\!\ast^1\!\ast^1$, type $\ast\ast/\ast\ast$; and $\alpha \to 1$, $P \to \langle 1 - t \rangle$, $Q \to \langle -t \rangle$, or $^1\!\ast^2\!\ast^2$, type $\ast\ast//\circ$.

$*^P\times^Z$ $1 = P^2 = (PZ^{-2}PZ^2)$. Again, this leads to two cases: $P \to 1$, $Z \to \langle t+1 \rangle$, or $*^1\times^p$, type $*\times/*\times$; and $P \to \langle 1-t \rangle$, $Z \to \langle -t \rangle$, or $*^2\times^1$, type $*\times//\circ$.

$\times^Y\times^Z$ $1 = Y^2Z^2$. There are two cases: $Y \to \langle t+1 \rangle$, $Z \to \langle t-1 \rangle$, or $x^p x^p$, type $\times\times/\times\times$; and $Y \to \langle 1-t \rangle$, $Z \to \langle -t \rangle$, or $\times^2\times^2$, type $\times\times//\circ$.

$\circ^{X,Y}$ $XY = YX$. Since the point group is trivial, all $p+1$ slopes are invariant. But, all are equivalent under automorphisms, so we obtain just one case: $X \to 1$, $Y \to \langle t+1 \rangle$, or $\circ^{1,p}$, type $\circ/\circ$.

Annotated Signature	Color Signature	Type	Restriction
$*^P 6^Q 3^R 2$	$*^1 6^{(12)} 3^{(13)} 2$	$*632^3/*333$	$p=3$
	$*^{(12)} 6^{(13)} 3^{(12)} 2$	$*632^3/2222$	$p=3$
$^\alpha 6^\beta 3^\gamma 2$	$^{(12)} 6^{(123)} 3^{(23)} 2$	$632^3/333$	$p=3$
	$^{(123)} 6^{(132)} 3^1 2$	$632^3/2222$	$p=3$
	$^6 6^3 3^2 2$	$632^p/\circ$	$p \equiv 1 \pmod 6$
$*^P 4^Q 4^R 2$	No cases		
$^\alpha 4*^P 2$	No cases		
$^\alpha 4^\beta 4^\gamma 2$	$^4 4^4 4^2 2$	$442^p/\circ$	$p \equiv 1 \pmod 4$
$*^P 3^Q 3^R 3$	$*^{(12)} 3^{(13)} 3^{(23)} 3$	$*333^3/\circ$	$p=3$
	$*^{(12)} 3^{(23)} 3^{(12)} 3$	$*333^3/333$	$p=3$
$^\alpha 3*^P 3$	$^{(123)} 3*^{(12)} 3$	$3*3^3/\circ$	$p=3$
	$3^{(123)} *^1 3$	$3*3^3/*333$	$p=3$
$^\alpha 3^\beta 3^\gamma 3$	$^1 3^{(123)} 3^{(132)} 3$	$333^3/333$	$p=3$
	$^{(123)} 3^{(123)} 3^{(123)} 3$	$333^3/\circ$	$p=3$
	$^3 3^3 3^3 3$	$333^p//\circ$	$p \equiv 1 \pmod 3$
$*^P 2^Q 2^R 2^S 2$	$*^1 2^2 2^1 2^2 2$	$*2222^p//**$	$p \equiv 1 \pmod 2$
$^\alpha 2*^P 2^Q 2$	$^2 2*^1 2^2 2$	$2*22^p//**$	$p \equiv 1 \pmod 2$
$^\alpha 2^\beta 2*^P$	$^2 2^2 2*^1$	$22*^p//**$	$p \equiv 1 \pmod 2$
	$^2 2^2 2^1 *^2$	$22*^p//\circ$	
$^\alpha 2^\beta 2\times^Z$	$^2 2^2 2\times^2$	$22\times^p//\times\times$	$p \equiv 1 \pmod 2$
$^\alpha 2^\beta 2^\gamma 2^\delta 2$	$^2 2^2 2^2 2^2 2$	$2222^p//\circ$	$p \equiv 1 \pmod 2$
$^\alpha_* *^P *^Q$	$^1_* *^2 *^2$	$**^p//\circ$	$p \equiv 1 \pmod 2$
	$^p_* *^1 *^1$	$**^p/**$	
$*^P \times^Z$	$*^2 \times^2$	$*\times^p//\circ$	$p \equiv 1 \pmod 2$
	$*^1 \times^p$	$*\times^p/*\times$	
$\times^Y \times^Z$	$\times^2 \times^2$	$\times\times^p//\circ$	$p \equiv 1 \pmod 2$
	$\times^p \times^p$	$\times\times^p/\times\times$	
$\circ^{X,Y}$	$\circ^{1,p}$	$\circ^p/\circ$	$p \equiv 1 \pmod 2$

Table 13.1. Primefold color types of plane patterns ($p \neq 2$.)

The Remaining Primefold Types for Plane Patterns

We have now found the threefold color versions of all 17 groups and indeed the p-fold ones for all except

$$*\mathbf{632}, \mathbf{632}, *\mathbf{442}, \mathbf{4{*}2}, \mathbf{442}, *\mathbf{333}, \mathbf{3{*}3}, \mathbf{333}.$$

We now close this gap, assuming $p \geq 5$, which ensures that p does not divide the order of the point group $\mathbf{Q}$. The simplest way to do so is to employ the Gaussian and Eisensteinian integers, two simple yet important sets of complex numbers that are not described in detail here.

The "Gaussian" Cases

In these cases,[3] the lattice may be thought of as the translations through Gaussian integers, $a + bi$ ($i = \sqrt{-1}$). Then the index-p sublattice must be closed under multiplication by i, since this is in $\mathbf{G}$; it is therefore an ideal of norm p.

Such things exist just when $p \equiv 1 \pmod 4$, and there are just two of them, consisting respectively of the multiples of $a + bi$ or $a - bi$ for some a and b with $a^2 + b^2 = p$. Neither of these is invariant under $*\mathbf{442}$ or $\mathbf{4{*}2}$ since those groups have symmetries that interchange $a + bi$ and $a - bi$. Since these same symmetries are automorphisms of $\mathbf{442}$ we have just one case: $^4\mathbf{4}^4\mathbf{4}^2\mathbf{2}$, type $\mathbf{442}^p/\circ$. (See Figure 13.2.)

Figure 13.2. The single case for $\mathbf{442}$, illustrated with $p = 5$ (and $a = 2, b = 1$).

[3]All three groups contain $\mathbf{442}$, which is generated by the maps $\alpha : z \to iz$ and $\gamma : z \to 1 - z$.

The "Eisensteinian" Cases

The discussion here is similar to the previous one except that we must replace the Gaussian integers by the Eisensteinian integers $a + b\omega$, where $\omega = \frac{1}{2}(-1 + \sqrt{-3})$ is a primitive cube root of 1. The index-p sublattice, being closed under multiplication by ω, is an ideal of norm p. Such things exist just when $p \equiv 1 \pmod 6$ and then consist of the multiples of $a + b\omega$ or $a + b\omega^2$ for some a and b with $a^2 + ab + b^2 = p$. Like those in the previous section, the groups $*\mathbf{632}$, $*\mathbf{333}$, and $\mathbf{3*3}$ have symmetries that interchange the two ideals. Again, these symmetries are automorphisms of $\mathbf{632}$ and $\mathbf{333}$, which therefore give just one case each: namely, $^6\mathbf{6}^3\mathbf{3}^2\mathbf{2}$, type $\mathbf{632}^p/\circ$, and $^3\mathbf{3}^3\mathbf{3}^3\mathbf{3}$, type $\mathbf{333}^p/\circ$. (See Figure 13.3.)

Figure 13.3.   The single cases for $\mathbf{632}$ and $\mathbf{333}$, illustrated with $p = 7$ (and $a = 2$, $b = 1$).

Spherical Patterns and Frieze Patterns

We can hope to p-fold color a spherical group only if p divides its order. Therefore, the only odd primes for which we can p-fold color the polyhedral groups are $p = 3$ or 5 for $*\mathbf{532}$ and $\mathbf{532}$ and $p = 3$ for $*\mathbf{432}$, $\mathbf{432}$, $*\mathbf{332}$, $\mathbf{3*2}$, and $\mathbf{332}$. It turns out that $*\mathbf{532}$ and $\mathbf{532}$ each have just one 5-fold coloring, shown below, but no 3-fold one, since neither has a subgroup of index 3.

In the remaining polyhedral groups, a subgroup of index 3 exists and is unique by Sylow's theorem, since it must be a Sylow-2-subgroup.

In any p-fold coloring (p odd) of one of the axial groups

$$*\mathbf{22N}, \mathbf{2*N}, \mathbf{22N}, *\mathbf{NN}, \mathbf{N*}, \mathbf{N\times}, \mathbf{NN},$$

we can suppose that the rotation of order N effects the permutation $(0, 1, 2, ..., p-1) = \langle t+1 \rangle$ since there must be some element that does so, and this is essentially the only choice. The coloring types that arised are listed in Table 13.2. Once again we merely telegraph the necessary arguments.

$*^P\mathbf{5}^Q\mathbf{3}^R\mathbf{2}$ $^\alpha\mathbf{5}^\beta\mathbf{3}^\gamma\mathbf{2}$ $*^P\mathbf{4}^Q\mathbf{3}^R\mathbf{2}$ $^\alpha\mathbf{4}^\beta\mathbf{3}^\gamma\mathbf{2}$

$*^P\mathbf{3}^Q\mathbf{3}^R\mathbf{2}$ $^\alpha\mathbf{3*}^P\mathbf{2}$ $^\alpha\mathbf{3}^\beta\mathbf{3}^\gamma\mathbf{2}$

$*^P\mathbf{5}^Q\mathbf{3}^R\mathbf{2}$ $1 = P^2 = (PQ)^5 = Q^2 = (QR)^3 = R^2 = (RP)^3$, abstractly $2 \times \mathbf{A}[5]$, has no subgroup of index 3 but one ($\cong 2 \times \mathbf{A}[4]$) of index 5: $*\mathbf{532}\,^5/\mathbf{3*2}/\times$.

$^\alpha\mathbf{5}^\beta\mathbf{3}^\gamma\mathbf{2}$ $\alpha\beta\gamma = 1 = \alpha^5 = \beta^3 = \gamma^2$, abstractly $\mathbf{A}[5]$, has no subgroup of index 3 but one ($\mathbf{A}[4]$) of index 5: $\mathbf{532}\,^5/\mathbf{332}/\mathbf{1}$.

$*^P\mathbf{4}^Q\mathbf{3}^R\mathbf{2}$ $1 = P^2 = (PQ)^4 = Q^2 = (QR)^3 = R^2 = (RP)^2$, abstractly $2 \times \mathbf{S}[4]$, has one 3-fold coloring given by Sylow's theorem: $*\mathbf{432}\,^3/*\mathbf{422}/*\mathbf{222}$.

$^{\alpha}\mathbf{4}^{\beta}\mathbf{3}^{\gamma}\mathbf{2}$ $\alpha\beta\gamma = 1 = \alpha^4 = \beta^3 = \gamma^2$ is abstractly $\mathbf{S}[4]$. Again, Sylow's theorem tells us there is only one 3-fold coloring: $^{(AB)}\mathbf{4}^{(ABC)}\mathbf{3}^{(AC)}\mathbf{2}$ or $^2\mathbf{4}^3\mathbf{3}^2\mathbf{2}$, $\mathbf{432}^3/\mathbf{422}/\mathbf{222}$.

$*^{P}\mathbf{3}^{Q}\mathbf{3}^{R}\mathbf{2}$ $1 = P^2 = (QP)^3 = Q^2 = (QR)^3 = R^2 = (RP)^2$ is also $\mathbf{S}[4]$ and has one 3-fold coloring: $*^{(AB)}\mathbf{3}^{(BC)}\mathbf{3}^{(AB)}\mathbf{2} = *^2\mathbf{3}^2\mathbf{3}^2\mathbf{2}$; $*\mathbf{332}^3/\mathbf{2}*\mathbf{2}/\mathbf{222}$.

$^{\alpha}\mathbf{3}*^{P}\mathbf{2}$ $1 = \alpha^3 = P^2 = \alpha P\alpha^{-1}P^{-1}$ is abstractly $2 \times \mathbf{A}[4]$ with one 3-fold coloring: $^{(ABC)}\mathbf{3}*^1\mathbf{2} = ^3\mathbf{3}*^1\mathbf{2}$.

$^{\alpha}\mathbf{3}^{\beta}\mathbf{3}^{\gamma}\mathbf{2}$ $\alpha\beta\gamma = 1 = \alpha^3 = \beta^3 = \gamma^2$ is simply $\mathbf{A}[4]$. It has one 3-fold coloring of type $\mathbf{332}^3/\mathbf{222}$.

We illustrate the remaining cases, taking $N = pM$ to be very large, with $p = 3$ in the figures. The analysis applies to the frieze groups $(N = \infty)$ as well.

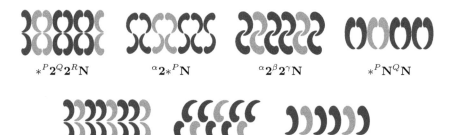

$*^{P}\mathbf{2}^{Q}\mathbf{2}^{R}\mathbf{N}$ $^{\alpha}\mathbf{2}*^{P}\mathbf{N}$ $^{\alpha}\mathbf{2}^{\beta}\mathbf{2}^{\gamma}\mathbf{N}$ $*^{P}\mathbf{N}^{Q}\mathbf{N}$

$^{\alpha}\mathbf{N}*^{P}$ $\mathbf{N}\times^{Z}$ $^{\alpha}\mathbf{N}\mathbf{N}$

$*^{P}\mathbf{2}^{Q}\mathbf{2}^{R}\mathbf{N}$ $1 = P^2 = (PQ)^2 = Q^2 = (QR)^2 = R^2 = (RP)^N$ has one p-coloring: $*^{\langle -t\rangle}\mathbf{2}^1\mathbf{2}^{\langle 1-t\rangle}\mathbf{2} = *^2\mathbf{2}^1\mathbf{2}^2\mathbf{2}$; $*\mathbf{22N}^p/*\mathbf{22M}/\mathbf{M}*$.

$^{\alpha}\mathbf{2}*^{P}\mathbf{N}$ $1 = \alpha^2 = P^2 = \alpha P\alpha^{-1}P^{-1}$ has one p-fold coloring: $^{\langle -t\rangle}\mathbf{2}*^{\langle 1-t\rangle}\mathbf{N}$; $\mathbf{2}*\mathbf{N}^p/\mathbf{2}*\mathbf{M}/\mathbf{M}\times$.

$^{\alpha}\mathbf{2}^{\beta}\mathbf{2}^{\gamma}\mathbf{N}$ $\alpha\beta\gamma = 1 = \alpha^2 = \beta^2 = \gamma^N$ has one such coloring: $^{\langle 1-t\rangle}\mathbf{2}^{\langle -t\rangle}\mathbf{2}^{\langle t+1\rangle}\mathbf{N} = ^2\mathbf{2}^2\mathbf{2}^p\mathbf{N}$; $\mathbf{22N}^p/\mathbf{22M}/\mathbf{MM}$.

$*^P \mathbf{N}^Q \mathbf{N}$ $1 = P^2 = (PQ)^N = Q^2$ has coloring $*^{\langle -t \rangle} \mathbf{N}^{\langle 1-t \rangle} \mathbf{N}$ or $*^p \mathbf{N}^p \mathbf{N}$; $*\mathbf{N}\mathbf{N}^p / *\mathbf{M}\mathbf{M}/\mathbf{M}\mathbf{M}$.

$^\alpha \mathbf{N}*^P$ $1 = \alpha^N = P^2 = \alpha P \alpha^{-1} P^{-1}$ has coloring $^p\mathbf{N}*^1$; $\mathbf{N}*^p/\mathbf{M}*$.

$\mathbf{N} \times^Z$ $1 = Z^{2N}$ has coloring $\mathbf{N}\times^p$; $\mathbf{N}\times^p/\mathbf{M}\times$.

$^\alpha \mathbf{N}\mathbf{N}$ $1 = \alpha^N$ has coloring $^p\mathbf{N}\mathbf{N}$; $\mathbf{N}\mathbf{N}^p/\mathbf{M}\mathbf{M}$.

Spherical Group	Color Type	Spherical Group	Color Type
$*\mathbf{532}$	$*\mathbf{532}^5//\times$	$*\mathbf{22N}$	$*\mathbf{22N}^p//\mathbf{M}*$
$\mathbf{532}$	$\mathbf{532}^5//\mathbf{1}$	$\mathbf{2}*\mathbf{N}$	$\mathbf{2}*\mathbf{N}^p//\mathbf{M}\times$
$*\mathbf{432}$	$*\mathbf{632}^3//*\mathbf{222}$	$\mathbf{22N}$	$\mathbf{2NN}^p//\mathbf{MM}$
$\mathbf{432}$	$\mathbf{432}^3//\mathbf{222}$	$*\mathbf{NN}$	$*\mathbf{NN}^p//\mathbf{MM}$
$*\mathbf{332}$	$*\mathbf{332}^3//\mathbf{222}$	$\mathbf{N}*$	$\mathbf{N}*^p/\mathbf{M}\times$
$\mathbf{3}*\mathbf{2}$	$\mathbf{3}*\mathbf{2}^3/*\mathbf{222}$	$\mathbf{N}\times$	$\mathbf{N}\times^p/\mathbf{M}\times$
$\mathbf{332}$	$\mathbf{332}^3/\mathbf{222}$	$\mathbf{NN}$	$\mathbf{NN}^p/\mathbf{MM}$

Table 13.2. Odd primefold color types of spherical and frieze patterns.

Where Are We?

Taken together, our "coloring chapters" have enumerated all prime-fold color types for repeating patterns on the plane and the sphere. While there are infinitely many further colorings to explore, we shall end our discussion here.

– 14 –

Searching for Relations

In this chapter, we establish the presentations that we introduced in Chapter 10. First, though, we must clear up the little matter of "left" and "right."

On Left and Right

If α and β are two group operations, what should $\alpha\beta$ mean? Many mathematicians write $\alpha(P)$ for the image of P under α, which has the rather confusing effect that $\alpha\beta$ means "β, followed by α". In this book, we adopt the more natural meaning "α followed by β" and so write $P\alpha$ for the image of P under α.

The underlying fact is that a group can act on its own elements in two different ways: either on the left or on the right. It is multiplication on the left that causes the confusing reversal: the result of multiplying α first by β and then by γ (on the left) is $\gamma\beta\alpha$ rather than $\alpha\beta\gamma$. We might avoid this problem by using *division* on the left, which gets the order right because dividing by α and then by β is equivalent to dividing by $\alpha\beta$—they both take γ to $\beta^{-1}\alpha^{-1}\gamma$.

Justifying the Presentations

One of the confusing things about geometrical groups is that they involve both left and right actions.

Our first figure shows the images of a particular point P under the group $^\alpha 6^\beta 3^\gamma 2$ generated by the particular rotations α, β, and γ shown below; note that $\alpha\beta\gamma = 1$.

(opposite page) The Cayley graph for our presentation of **632.**

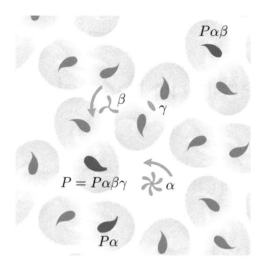

It is annoying that the image $P\alpha\beta$ of a typical point $P\alpha$ appears far away in the figure from $P\alpha$. In other words, it is hard to see the action by right multiplication. The left action, although it has less geometrical meaning, is much easier!

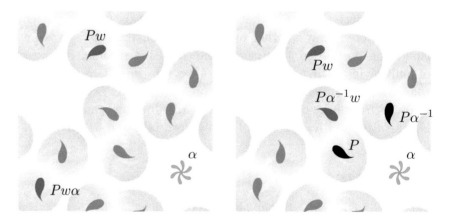

It is annoying that, after applying α to Pw, the image $Pw\alpha$ is quite far away. On the other hand $P\alpha^{-1}w$ is just as close to Pw as $P\alpha^{-1}$ is to Pw.

For the reasons given, we actually use left division, rather than left multiplication. For any word w, the points Pw and $P\alpha^{-1}w$, being the images of P and $P\alpha^{-1}$ under w, are just as close to each other as those two points. So the green arrows, which lead from each point Pw to the corresponding $P\alpha^{-1}w$ in the figure on page 170, are all the same length. Of course, the same is true of the red $(Pw \to P\beta^{-1}w)$ and yellow $(Pw \to P\gamma^{-1}w)$ arrows.

Since left division is a homomorphism, our relations $\alpha\beta\gamma = 1 = \alpha^6 = \beta^3 = \gamma^2$ must hold in this diagram. This is why the green arrows form hexagons (because $\alpha^6 = 1$), the red ones triangles ($\beta^3 = 1$), and the yellow ones digons ($\gamma^2 = 1$), while following green-then-red-then-yellow arrows always determines a triangle, because $\alpha\beta\gamma = 1$.

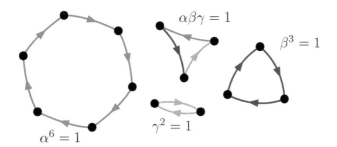

These tiles correspond to the relations.

The Sufficiency of the Relations

Because the relations describe the borders of tiles, it is easy to see why they suffice. The argument uses the fact that the polygons corresponding to the relators tile the plane, which topologically is a *simply-connected* manifold. This means that any closed path in the plane can be shrunk to a point without leaving the plane. This is also true for the sphere but is false for the torus. Now suppose that the geometrical transformation corresponding to some complicated product (or "word") w in α, β, γ is the identity. Then, in particular, we must have $Pw^{-1} = P$, which shows that the path starting at P that corresponds to this word must return to P.

One such word is $(\alpha^2\beta^{-1}\gamma^{-1})^2$, which corresponds to the blue path in the next figure. Since this is closed, it can be shrunk to a point!

Simple-Connectedness

The sphere is simply connected—every closed path can be shrink to a point—as is the plane, but the torus is not. Though some closed paths on the torus can be shrunk to a point, some cannot.

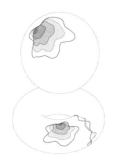

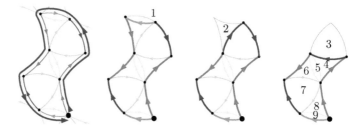

We can do this by a sequence of alterations, each of which replaces a path partway around some tile by the remainder of the boundary of that tile. For example, cutting off the digon 1 replaces the first γ^{-1} by γ, leading to $\alpha^2\beta^{-1}\gamma\alpha^2\beta^{-1}\gamma^{-1}$; then, cutting off triangle 2 will replace $\gamma\alpha$ by β, leading to $\alpha^2\beta^{-2}\alpha\beta^{-1}\gamma^{-1}$. These are the first two of the nine replacements that we suggest in figures above, the next being that of β^{-2} by β (triangle 3) and the last (using digon 9) that of γ^2 (which is all that then remains of the word).

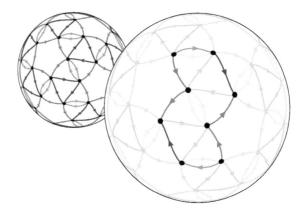

The same argument would establish the analogous presentation

$$^\alpha 5^\beta 3^\gamma 2 : \alpha\beta\gamma = 1 = \alpha^5 = \beta^3 = \gamma^2$$

for **532**, even though the analogous figure for that case is on a sphere rather than the Euclidean plane. This makes no difference because the sphere is still a simply-connected manifold—any closed loop can still be shrunk to a point on it.

The General Case

The argument for **632** worked because our figure, which was a "Cayley graph" for that group, was formed by the edges of a particular kind of tiling. Namely, the vertices of the tiling are in exact correspondence with the elements of the group, there is one type of directed edge for each generator, and each type and direction of edge appears at each vertex. One vertex, g, is connected to another, h, by an edge corresponding to a generator α if and only if $h = g\alpha$.

If a generator is its own inverse we draw a doubled edge instead of two directed edges.

There is one type of tile for each relator; the relator corresponding to any tile is the word obtained by reading the generators around that tile. Of course, for our two-dimensional groups, each edge must belong to just two tiles.

Loops in the tiling correspond to words that represent the identity, simply because we may shrink such a loop to a point, just as we saw for **632** and **532**.

We call this a *van Kampen tiling*. The justification of our presentations for the other groups reduces to finding analogous van Kampen tilings.

Recall that the orbifold of any of our groups is obtained from the sphere by adding some handles, cone points, boundary curves with corners, and crosscaps. We will carve this up into pieces that "lift up" to a van Kampen tiling when we unfold the orbifold.

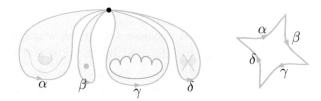

The global relation: We fix a basepoint and carve the orbifold up into pieces that will lift to the tiles of our van Kampen tiling. First, we separate off each piece of the orbifold with a loop from the base point. The portion that remains is a disk corresponding to the global relation. Here, $\alpha\beta\gamma\delta = 1$.

The cutting paths shown in the following figures represent the elements of the fundamental group of this orbifold that correspond to our chosen generators.

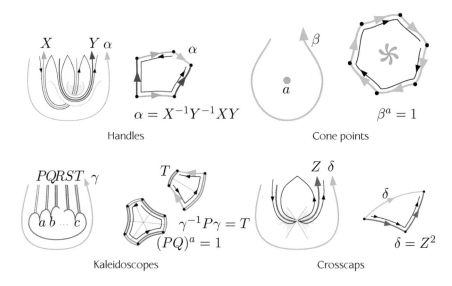

Handles: We can carve a handle to lie flat by cutting along two curves X and Y, producing a tile corresponding to the relation $\alpha = X^{-1}Y^{-1}XY$.

Cone points: The loop around a cone point of order A already separates off a piece that lifts to a tile, corresponding to a relation $\beta^A = 1$.

Kaleidoscopes: So that the pieces will lift to tiles that are disks, we must carve apart the kaleidoscopic points of a kaleidoscope, obtaining the relations $P^2 \ldots, T^2, (PQ)^a = \ldots = (ST)^d = 1$, and $\gamma^{-1}P\gamma = T$.

Crosscaps: A crosscap can be cut open along a single loop to produce a flat tile, corresponding to the relation $Z^2 = \delta$.

The way we get the desired tiling is very simple: we just "lift" the figure to the original surface by unfolding the orbifold! We illustrate this with the presentation

$$^\alpha 4^\beta *^P 2^Q : \alpha\beta = 1 = \alpha^4 = P^2 = (PQ)^2 = Q^2, Q = \beta^{-1}P\beta$$

for the group **4∗2**. The orbifold of this group is very simple: it looks like a sort of ice-cream cone, but we redraw it as at the right below to show it as part of the general case.

The figure below shows the unfolded tiling that corresponds to this. The reader should check the essential properties that each generator appears once inwards and once outwards at each vertex and that the word around each tile is a relator. The argument we used for **632** now shows that **4∗2** does indeed have the given presentation.

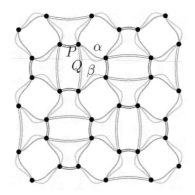

The previous diagram contains three types of two-sided region. We can simplify it by collapsing these to single edges, as in the figure below. Identifying β with α^{-1} collapses the pink regions, while noticing that P and Q are self-inverse does the same for the two green types of two-sided region.

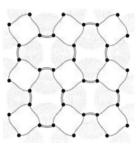

We shall routinely collapse two-sided regions in this way. However, there can be less obvious simplifications. For instance, omitting all the edges of one particular kind sometimes still yields a tiling with the van Kampen properties, which then corresponds to a presentation obtained by making the appropriate substitution for the corresponding generator. This happens for either of the two types of green edge in our example. We may eliminate all the edges of a particular kind, so long as our graph remains connected.

Simplifications

Nothing more needs to be said to justify the presentation in the general case, but it is interesting to observe how the form of the tiling simplifies in various special cases. Thus, the generic presentation $^\alpha *^P 6^Q 3^R 2^S$ for $*\mathbf{632}$ is shown in the next figure.

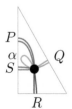

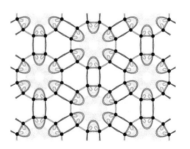

We can eliminate the "peapods" whose two "peas" are one-sided faces corresponding to the relation $\alpha = 1$ contained in a four-sided face that completes the "pod" and corresponds to the relation $\alpha^{-1}P\alpha = S$. We obtain another simplification by dropping α and identifying S with P, which is equivalent to collapsing these pods. The resulting diagram is typical of those for kaleidoscopic groups.

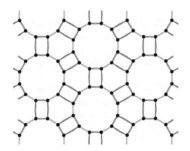

The justification of our presentations is now complete, but we add a final note for the interested reader on why the argument takes the form that it does.

Alias and Alibi

The real problem is, why do the left and right actions of G both occur here? The reason is that a permutation—say, of the names of a number of people—can be thought of as moving either the names or the people. The alias viewpoint regards the permutation as assigning a new name or "alias" to each person (from the Latin *alias* = otherwise). Alternatively, from the alibi viewpoint we move the people to the places corresponding to their new names (from the Latin *alibi* = in another place).

The two resulting mathematical descriptions of the permutation are inverses:

Jane moves to John's spot,
John to Baby's,
and Baby to Jane's,

Figure 14.1. On the left, changing each alias; on the right, changing alibis. The effect is the same, but expressed as group operations, one is Jane→John→Baby→Jane, the other Jane→Baby→John→Jane.

but

> Jane's spot becomes Baby's,
> Baby's becomes John's,
> and John's becomes Jane's.

So if we use right multiplication for one of them, we should use left multiplication for the other.

In the figures at the beginning of this chapter, we named various points (Pw) by group elements (w). The group acts on points by multiplication on the right, and so it is natural that it should act on the left for their names or, equivalently, for the group elements. It is the action on group elements that was needed to understand the relations.

Where Are We?

We have now justified the presentations that we used to enumerate coloring types in Chapters 11–13.

Exercises

Now for an exercise: name the symmetries of the following graphs, and decipher the presentations. One group has two graphs shown, corresponding to two different presentations discussed in Chapter 10.

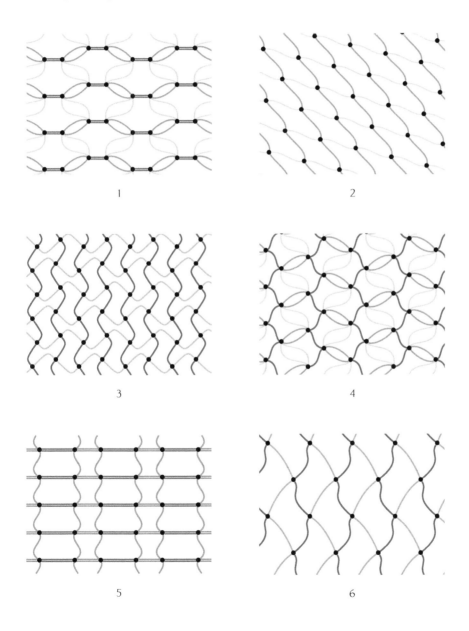

1

2

3

4

5

6

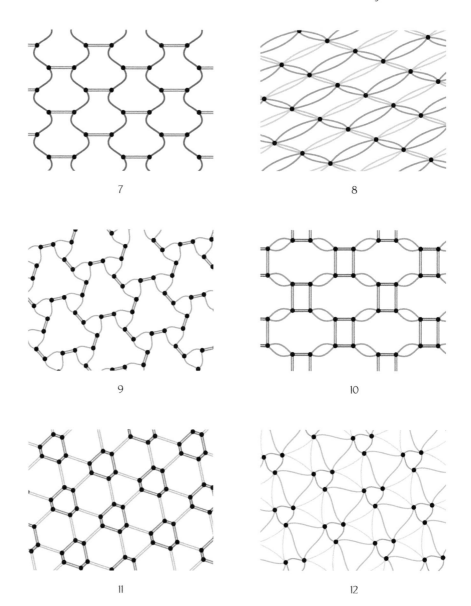

7

8

9

10

11

12

Answers to Exercises

1. **22**∗; $1 = \alpha^2 = \beta^2 = P^2, P\alpha\beta = \alpha\beta P.$

2. ○; $XY = YX.$

3. $\mathbf{22}\times$; $1 = (YZ)^2 = (YZ^{-1})^2$.

4. $\mathbf{22}\times$; $1 = \alpha^2 = \beta^2 = \infty BZ^2$.

5. $**$; $1 = P^2 = Q^2, \alpha P = P\alpha, \alpha Q = Q\alpha$.

6. $\times\times$; $1 = Y^2 Z^2$.

7. $*\times$; $1 = P^2, PZ^2 = Z^2 P$.

8. $\mathbf{2222}$; $\alpha\beta\gamma\delta = 1 = \alpha^2 = \beta^2 = \gamma^2 = \delta^2$.

9. $\mathbf{3*3}$; $1 = P^2 = \alpha^3 = (\alpha^{-1} P \alpha P)^3$.

10. $\mathbf{2*22}$; $1 = P^2 = Q^2 = (PQ)^2 = (\alpha^{-1} P \alpha Q)^2$.

11. $*\mathbf{333}$; $1 = P^2 = (PQ)^3 = Q^2 = (QR)^3 = R^2 = (RP)^3$.

12. $\mathbf{333}$; $\alpha\beta\gamma = 1 = \alpha^3 = \beta^3 = \gamma^3$.

– 15 –

Types of Tilings

Heesch Types

In how many topologically different ways can we choose a simply connected fundamental domain ("the tile") for one of the seventeen plane groups? This subject was investigated by Heesch and independently by Escher, and an exhaustive enumeration of the plane types is found in Grünbaum and Shephard [18].

The very much simpler treatment we give here was pioneered by Daniel Huson and Olaf Delgado. The answer, of course, depends on the group—obviously for a reflection group there is just one fundamental domain. On the other hand, there are four for **632** as shown on the next page. Why is this? The answer is found by looking at the orbifold: a graph on the orbifold will be the boundary of a fundamental domain if it cuts the orbifold into a topological disk that has no internal cone point and can be opened flat onto the plane. We just call such a disk a *flat disk*. There are two topologically distinct ways to cut open a sphere with three cone points into such a tile, shown below.

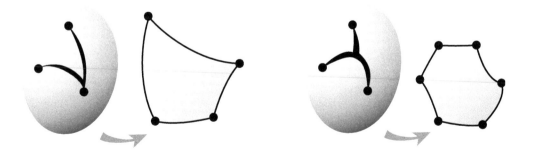

(opposite page) A Heesch tiling made by unusual paving stones, in Zakopane, Poland.

185

The previous figure is an example of the visual shorthand that we use to represent various types of orbifolds. Here is the complete list:

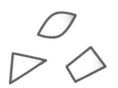

Spheres with cone points: signatures **NN**, **22N**, **332**, **333**, **432**, **442**, **532**, **632**, **2222**.

Disks with kaleidoscopic points: signatures **∗NN**, **∗22N**, **∗332**, **∗333**, **∗432**, **∗442**, **∗532**, **∗632**, **∗2222**.

Disks with kaleidoscopic points and cone points: **2∗N**, **2∗22**, **3∗2**, **3∗3**, **4∗2**.

Disks with cone points: **N∗**, **22∗**.

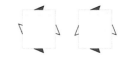

Crosscaps with cone points: **N×**, **22×**.

The twice-punched sphere, or annulus, **∗∗**; and the once-punched crosscap, or Möbius band, **∗×**.

A pair of crosscaps, or klein bottle, **× ×**; and a handle, or torus, **○**.

Now back to a sphere with three cone points. If the cone points have distinct orders, as for **632**, we get four possible graphs altogether: we can label the cone points of the second graph in just one way, topologically, but the other can be labeled in three distinct ways. So there are four possible fundamental domains.

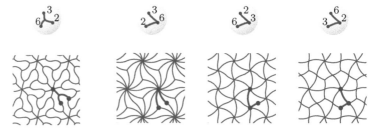

For **4∗2**, there are just two graphs possible on the orbifold, and so just two possible fundamental domains. The reason a reflection group has a unique fundamental domain is that its orbifold is already a flat disk!

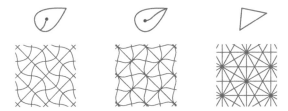

Groups with topologically equivalent orbifolds have corresponding kinds of fundamental domains since they will have topologically
the same graphs; this holds whether the types are on the sphere or
the plane. For example, the types **4∗2**, **3∗3**, **3∗2**, and **2∗N** have
topologically equivalent orbifolds: a disk with one cone point in the
interior and one kaleidoscopic point on the boundary. Below are
tilings of several types, arising from (topologically) the same graph
on (topologically) the same orbifold.

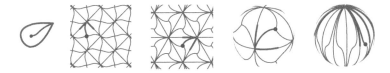

The figures on the following pages show the answers for all the
planar and spherical types. They were found by enumerating graphs
on the corresponding orbifolds that cut it into a disk and pass through
all its cone points.

Isohedral Types

Grünbaum and Shephard also list all types of *isohedral* tilings—
which is to say, those whose tiles are all of the same type in the
precise sense that there are symmetries taking any tile to all the
others. Since the symmetries that fix a tile form at most a point
group **N·** or ∗**N·**, the condition for this is that the graph must cut
the orbifold into a topological disk that has, at worst, either a single
cone point on its boundary or a single cone point in its interior. It
may include parts of the boundary.

For example, four such graphs are possible on the orbifold for **4∗2**, yielding four new isohedral tilings in addition to the two Heesch tilings already shown.

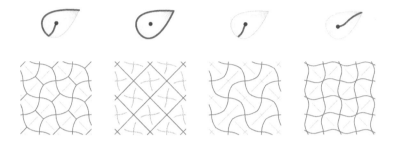

As before, the topology of the orbifold is what matters. Below are isohedral tilings with symmetry types **4∗2**, **3∗3**, **3∗2**, and **2∗N**, all arising from the same kind of graph on the same kind of orbifold.

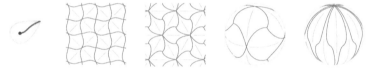

There may be symmetries of the tile itself that do not extend to the tiling. We indicate this possibility, for it is clearly important to designers and other tile users.

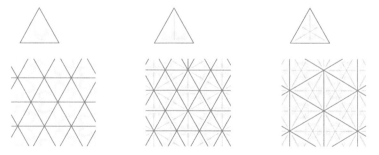

The types of isohedral tiling for the plane groups are shown below and on the following pages.

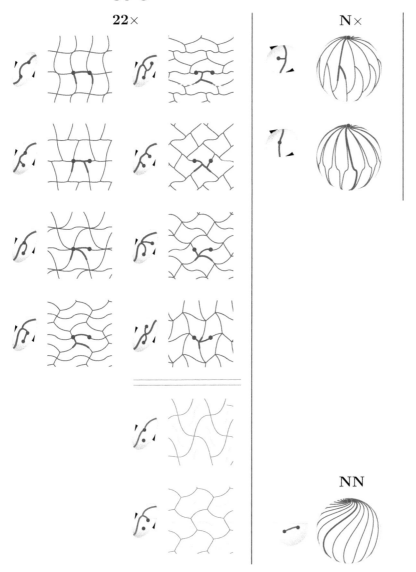

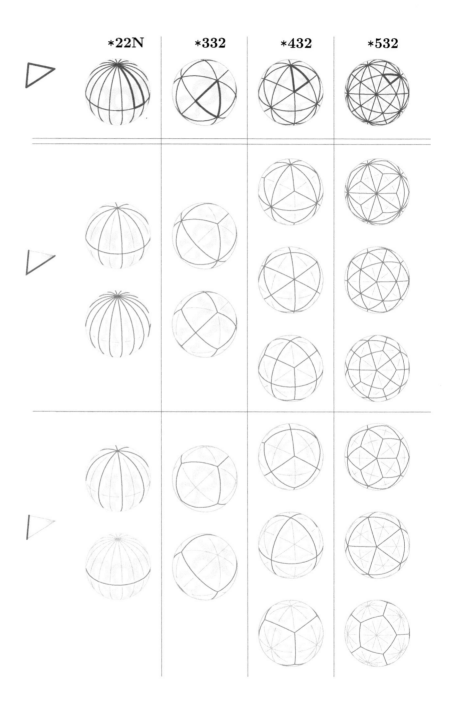

*632   *442   *333   *2222

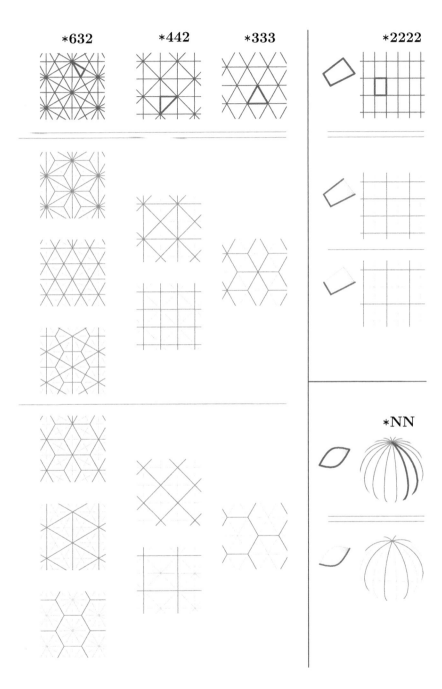

*NN

632   **442**   **333**   **2222**

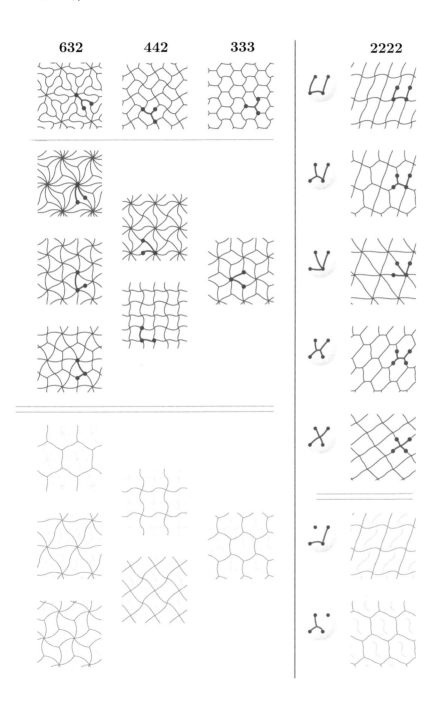

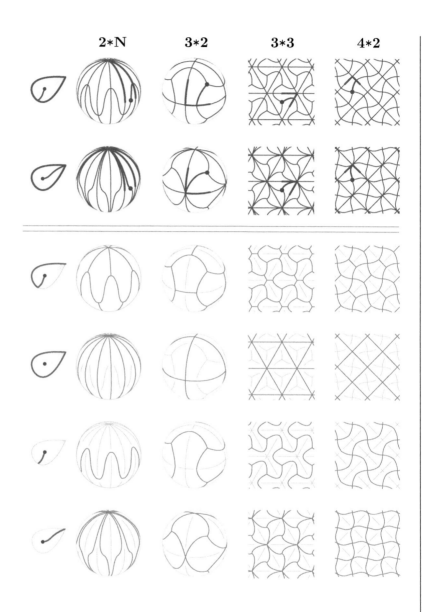

2∗22   22∗   N∗

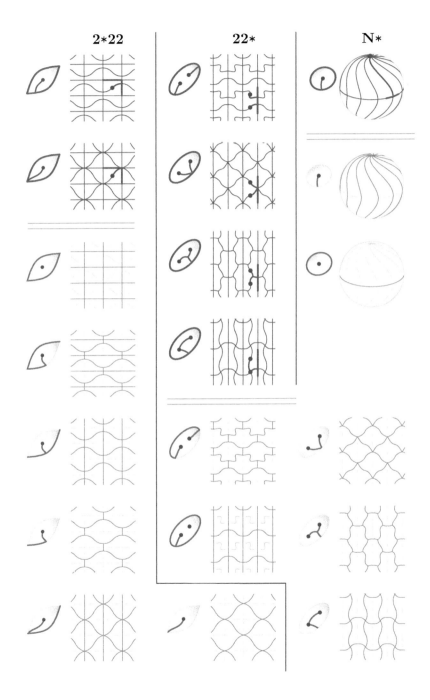

**

*×

××

○

Where Are We?

In this chapter, we used our orbifold theory to enumerate the isohedral tilings of the sphere or plane (the so-called Heesch types).

Order	(Number)	Groups
1	(1)	1.
2	(1)	2.
3	(1)	3.
4	(2)	$4, 2 \times 2$.
5	(1)	5.
6	(2)	$6 = 3 \times 2, \mathbf{D}_6 = 3 : 2 = \mathbf{S}[3]$.
7	(1)	7.
8	(5)	$8, 4 \times 2, 2 \times 2 \times 2, \mathbf{D}_8 = 4 : 2 = 2^{1+2}_+, \mathbf{Q}_8 = 4|2 = 2^{1+2}_-$.
9	(2)	$9, 3 \times 3$.
10	(2)	$10 = 5 \times 2, \mathbf{D}_{10} = 5 : 2$.
11	(1)	11.
12	(5)	$12 = 3 \times 4, 3 \times 2 \times 2, \mathbf{D}_{12} = 6 : 2, \mathbf{Q}_{12} = 6|2, \mathbf{A}[4] = (2 \times 2) : 3$
13	(1)	13.
14	(2)	$14 = 7 \times 2, \mathbf{D}_{14} = 7 : 2$.
15	(1)	$15 = 5 \times 3$.
16	(14)	$16, 8 \times 2, 4 \times 4, 4 \times 2 \times 2, 2 \times 2 \times 2 \times 2,$ $2 \times \mathbf{D}_8, 2 \times \mathbf{Q}_8, 2^{1+3} = 4 \circ \mathbf{Q}_8 = 4 \circ \mathbf{D}_8, \mathbf{D}_{16}, \mathbf{Q}_{16},$ $\mathbf{QD}_{16} = 8 : 2_3, 8 : 2 = 8 : 2_5, 4 : 4, (2 \times 2) : 4$.
17	(1)	17.
18	(5)	$18 = 9 \times 2, 3 \times 3 \times 2, 3 \times \mathbf{D}_6, \mathbf{D}_{18}, (3 \times 3) : 2 = (3 \times 3) : 2_{-1}$.
19	(1)	19.
20	(5)	$20 = 5 \times 4, 5 \times 2 \times 2, \mathbf{D}_{20} = \mathbf{D}_{10} \times 2, \mathbf{Q}_{20}, 5 : 4 = 5 : 4_2$.
21	(2)	$21 = 7 \times 3, 7 : 3$.
22	(2)	$22 = 11 \times 2, \mathbf{D}_{22} = 11 : 2$.
23	(1)	23.
24	(15)	$24 = 3 \times 8, 3 \times 4 \times 2, 3 \times 2 \times 2 \times 2, 3 \times \mathbf{D}_8, 3 \times \mathbf{Q}_8,$ $4 \times \mathbf{D}_6, 2 \times 2 \times \mathbf{D}_6 = 2 \times \mathbf{D}_{12}, 2 \times \mathbf{Q}_{12}, 3 : \mathbf{D}_8, 3 : \mathbf{Q}_8,$ $\mathbf{D}_{24}, \mathbf{Q}_{24}, \mathbf{S}[4] = (2 \times 2) : \mathbf{S}[3], 2 \times \mathbf{A}[4], 2|\mathbf{A}[4] = \mathbf{Q}_8 : 3$.
25	(2)	$25, 5 \times 5$.
26	(2)	$26 = 13 \times 2, \mathbf{D}_{26} = 13 : 2$.
27	(5)	$27, 9 \times 3, 3 \times 3 \times 3, 3^{1+2}_+ = (3 \times 3) : 3, 3^{1+2}_- = 9 : 3$.
28	(4)	$28 = 7 \times 4, 7 \times 2 \times 2, \mathbf{D}_{28} = 2 \times \mathbf{D}_{14}, \mathbf{Q}_{28}, 7 : 4$.
29	(1)	29.
30	(4)	$30 = 5 \times 3 \times 2, 5 \times \mathbf{D}_6, 3 \times \mathbf{D}_{10}, \mathbf{D}_{30}$.
31	(1)	31.

– 16 –

Abstract Groups

It seems silly to have enumerated so many different kinds of geometrical groups without enumerating abstract groups in their own right. We now rectify this omission.

Hans Ulrich Besche, Bettina Eick, and Eamonn O'Brien have celebrated the completion of the second millenium by publishing a paper [1] that gives the number of groups of each order up to 2000. The table that forms the appendix to this chapter, taken from their paper with their help and permission, extends this to 2009 so as to contain the publication date of this book.

But what about the groups themselves? The frontispiece gives our short names for the 93 groups whose order is at most 31. We briefly explain the notations used.

Cyclic Groups, Direct Products, and Abelian Groups

A number N represents the cyclic group (often also called $\mathbf{C}_N$) of order N. We use $\mathbf{G} \times \mathbf{H}$ for the direct product of groups $\mathbf{G}$ and $\mathbf{H}$ — if $\mathbf{G} = \langle g_1, g_2, \ldots | r_1, r_2, \ldots \rangle$ and $\mathbf{H} = \langle h_1, h_2, \ldots | s_1, s_2, \ldots \rangle$ are presentations for $\mathbf{G}$ and $\mathbf{H}$, then one for $\mathbf{G} \times \mathbf{H}$ is

$$\mathbf{G} \times \mathbf{H} = \langle g_1 \ldots, h_1 \ldots | r_1, r_2, \ldots, s_1, s_2, \ldots, g_i h_j = h_j g_i \rangle,$$

it being supposed that the generators for $\mathbf{G}$ and $\mathbf{H}$ don't overlap. It is a well-known theorem that every finite abelian group is a direct product of cyclic groups.

(opposite page) The groups of orders 1 through 31.

Split and Non-split Extensions

We say that $\mathbf{G}$ is an *extension*[1] of $\mathbf{H}$ by $\mathbf{K}$ and write $\mathbf{G} = \mathbf{HK}$ or $\mathbf{H.K}$ to mean that $\mathbf{G}$ has a normal subgroup $\mathbf{H}$ whose quotient $\mathbf{G}/\mathbf{H}$ is isomorphic to $\mathbf{K}$. The extension is a *split* one, written $\mathbf{G} = \mathbf{H} : \mathbf{K}$, if it has a subgroup disjoint from $\mathbf{H}$ and isomorphic to $\mathbf{K}$, and otherwise *non-split*, written $\mathbf{G} = \mathbf{H}|\mathbf{K}$. Usually we still need more information to specify the structure of $\mathbf{G}$. If $\mathbf{K} = \mathbf{N}$ is a cyclic group of order N generated by c, then it is enough to specify the automorphism γ of $\mathbf{H}$ defined by $c^{-1}hc = h^\gamma$. We do this by writing $\mathbf{G} = \mathbf{H} : \mathbf{N}_\gamma$, where the subscript may either name γ completely or just hint at it—for example a number n denotes the nth power map $h \mapsto h^n$. Often we omit the hint completely, understanding that $\mathbf{G} : \mathbf{N}$ means the only group of form $\mathbf{G} : \mathbf{N}_\gamma$ for which γ has as large an order as possible subject to having no simpler name.

Non-split extensions by cyclic groups are named similarly, replacing : by |, except that we might also need to specify the element $z = c^N$. We write $\mathbf{G} = \mathbf{H}|_z\mathbf{N}_\gamma$ when $c^{-1}hc = h^\gamma$ and $c^N = z$, but since z is often determined by the condition that it must be a central element of $\mathbf{H}$ that is fixed by γ, we can usually omit it.

Some of the simplest split and non-split extensions are the dihedral and quaternionic groups defined in the next section. The more interesting ones in the table are $\mathbf{A}[4] = (2 \times 2) : 3$ in which the automorphism cyclically permutes the three nontrivial elements of 2×2; $\mathbf{S}[4] = (2 \times 2) : \mathbf{S}[3]$, where the $\mathbf{S}[3]$ achieves all permutations of these elements; and $2|\mathbf{A}[4]$, which is simpler to describe as $\mathbf{Q}_8 : 3$, in which the automorphism is the one described below.

Dihedral, Quaternionic, and QuasiDihedral Groups

Three closely related groups have long received special names. They are $n : 2_{-1}$, the *Dihedral* group $\mathbf{D}_{2n}$ of order $2n$; $2n|2_{-1}$, the *Quaternionic* group $\mathbf{Q}_{4n}$ of order $4n$; and $2n : 2_{n-1}$, the *QuasiDihedral* group $\mathbf{QD}_{4n}$ of order $4n$.

---

[1]Since there are two conventions, more precisely this is an *upward* extension of $\mathbf{H}$ by $\mathbf{K}$ or a *downward* extension of $\mathbf{K}$ by $\mathbf{H}$.

The group $\mathbf{Q}_8$ has been notorious as the group of the unit quaternions $\pm 1, \pm i, \pm j, \pm k$ ever since October 16, 1843, when Hamilton carved its defining relations in the form

$$i^2 = j^2 = k^2 = ijk = -1$$

into the stone of Dublin's Brougham bridge. It has an obvious automorphism cyclically permuting i, j, k.

The two nonabelian groups $\mathbf{D}_8$ and $\mathbf{Q}_8$ of order 8 are very similar, and lead us into a new topic.

Extraspecial and Special Groups

An *extraspecial p-group* is one with a center of prime order p that is also its derived group and in which the pth power of every element is in the center. (The derived group of a group is the subgroup generated by commutators $a^{-1}b^{-1}ab$.) It follows from this description that any extraspecial group has structure

$$p \mid \quad \underbrace{(p \times p \times \ldots p)}_{e} \quad , \text{ or } p^{1+e} \text{ in brief.}$$

It turns out that $e = 2n$ is always even and that there are two cases p_+^{1+2n} and p_-^{1+2n} for each n. Moreover, these cases are central products

$$p_\pm^{1+2n} = \underbrace{\mathbf{X}_\pm \circ \mathbf{X}_+ \circ \mathbf{X}_+ \cdots}_{n}$$

of the two nonabelian groups of order p^3, say $\mathbf{X}_+ = p_+^{1+2}$ and $\mathbf{X}_- = p_-^{1+2}$. The *central product* $\mathbf{G} \circ \mathbf{H}$ of two groups is defined only when they have isomorphic central subgroups, say $\mathbf{Z}_1$ and $\mathbf{Z}_2$; then it is obtained from $\mathbf{G} \times \mathbf{H}$ by adding the relations that equate $\mathbf{Z}_1$ to $\mathbf{Z}_2$.

To complete this description, we need only identify $\mathbf{X}_+$ and $\mathbf{X}_-$. They are

$$2_+^{1+2} = \mathbf{D}_8 \text{ and } 2_-^{1+2} = \mathbf{Q}_8, \text{ when } p = 2,$$
$$p_+^{1+2} = (p \times p) : p \text{ and } p_-^{1+2} = p^2 : p, \text{ for odd } p,$$

where the implied automorphism takes $a \mapsto ab, b \mapsto b$ in the first case and $a \mapsto a^{p+1}$ in the second. In fact, any central product of factors $\mathbf{X}_\pm$ is extraspecial, but since $\mathbf{X}_- \circ \mathbf{X}_-$ is isomorphic to $\mathbf{X}_+ \circ \mathbf{X}_+$ (if $p = 2$) or $\mathbf{X}_+ \circ \mathbf{X}_-$ (if p is odd), we can suppose there is at most one $-$ sign.

Closely related to the two extraspecial groups $p_\pm^{1+e}$ $(e = 2n)$ is the *special* group p^{1+d} $(d = 2n + 1)$, defined by

$$p^{1+d} = \underbrace{\mathbf{X}_\pm \circ \mathbf{X}_\pm \circ \ldots \mathbf{X}_\pm}_{n} \circ \mathbf{C} \quad ,$$

where $\mathbf{C}$ is the cyclic group of order p^2 and the common central subgroup has order p. The signs are irrelevant since $\mathbf{X}_+ \circ \mathbf{C}$ and $\mathbf{X}_- \circ \mathbf{C}$ are isomorphic.

Groups of the Simplest Orders

We have now said enough to identify all the groups listed on page 196. The rest of the chapter is about groups whose order factorizes in some simple way.

p-Groups

Groups whose orders are powers of a prime p are normally called p-groups: they were enumerated through order p^4 by Hölder in 1893, and we shall briefly describe them in this section.

For $p = 2$ they appear on page 196 and for p odd in Table 16.1, except that for order 3^4 there are certain changes. For $p \geq 3$, all groups in a given line have the same number of elements of every order as the first one, which is an abelian group.

We can take the implied automorphisms to be

$$
\begin{aligned}
&\text{for } p^3 : p & &a \mapsto a^{p^2+1} \\
&\quad\;\; p^2 : p^2 & &a \mapsto a^{p+1} \\
&\quad\;\; (p \times p) : p^2 & &a \mapsto ab, b \mapsto b \\
&\quad\;\; (p \times p \times p) : \text{ or } |p & &a \mapsto ab, b \mapsto bc, c \mapsto c \\
&\quad\;\; (p^2 \times p) : p_\pm & &a \mapsto ab^i, b \mapsto a^{pj}b, \text{ where } \left(\tfrac{ij}{p}\right) = \pm 1.
\end{aligned}
$$

Order	(Number)	Groups
p	(1)	p.
p^2	(2)	p^2, $p \times p$.
p^3	(5)	p^3, $p^2 \times p$; $p^2 : p = p_-^{1+2}$, $p \times p \times p$; $(p \times p) : p = p_+^{1+2}$.
p^4 $(p > 3)$	(15)	p^4, $p^3 \times p$; $p^3 : p$, $p^2 \times p^2$; $p^2 | p^2$, $p^2 \times p \times p$; $(p \times p) : p^2$, $p \times p_-^{1+2}$, p^{1+3}; $\quad (p^2 \times p) : p_+$, $(p^2 \times p) : p_-$, $(p \times p \times p)|p$, $p \times p \times p \times p$; $p \times p_+^{1+2}$, $(p \times p \times p) : p$.

Table 16.1. The p-groups.

The two groups $(p^2 \times p) : p_\pm$ are very similar and have been called *the twins*. They are distinguished only by the quadratic residuacity of ij (mod p) — the exact values of i and j being otherwise unimportant. For $(p \times p \times p)|p$ the pth power of the automorphism may be taken to be c. However, for this and the earlier groups we don't really need to specify the automorphism since all appropriate values for it lead to the same group.

For the groups of order 3^4, one should make the following changes: in the line starting $9 \times 3 \times 3$ there is no group of the form $3^3|3$ — instead in line 9×9 we have a group $(9 \times 3)|3$ for which the automorphism is the same as in $(9 \times 3) : 3_-$. The numbers of elements of orders $1, 3, 9$ in $(9 \times 3) : 3_-$ and $(3 \times 3 \times 3) : 3$ are $1, 62, 18$ and $1, 44, 36$.

Groups of Squarefree Order

Hölder also classified all the groups of squarefree orders. We shall describe them as follows. For any particular group of *squarefree* order $pqr \ldots$, there are generators whose orders are the primes $p, q, r, \ldots$. The further relations are all of the form either $xy = yx$ or $y^{-1}xy = x^k \neq x$. If there is a relation of form $y^{-1}xy = x^k \neq x$, we say that y *acts* on x and call y an *actor* and x a *reactor*. No generator may be both an actor and a reactor, but there may be generators that

Opportunity Graph	(Number)	Groups
p	(1)	p

Opportunity Graph	(Number)	Groups
$p \quad q$	(1)	$p \times q$
$p \leftarrow q$	(2)	$p \times q,\ (p : q)$

Table 16.2. Cases p and pq.

Opportunity Graph	Shape name	(Number)	Groups
q $p \qquad r$	0	(1)	$p \times q \times r$
q $\swarrow$ $p \qquad r$	1	(2)	$p \times q \times r,\ (p : q) \times r$
q $\swarrow \ \nwarrow$ $p \qquad r$	2 seq	(3)	$p \times q \times r,\ (p : q) \times r,\ p \times (q : r)$
q $\swarrow$ $p \ \leftarrow \ r$	2 in	(4)	$p \times q \times r,\ (p : q) \times r, (p : r) \times q,$ $p : (q \times r)$
q $\nwarrow$ $p \ \leftarrow \ r$	2 out	$(r{+}2)$	$p \times q \times r,\ p \times (q : r),\ q \times (p : r),$ $(p \times q) : r_{i,j}$
q $\swarrow \ \nwarrow$ $p \ \leftarrow \ r$	3	$(r{+}4)$	$p \times q \times r,\ (p : q) \times r,$ $p \times (q : r),\ q \times (p : r),$ $p : (q \times r)\ ,\ (p \times q) : r_{i,j}$

Table 16.3. Case pqr.

are neither; such generators z we may call *commuters*, since they commute with all the other generators.

A generator y of order q can only act on x of order p if $p \equiv 1 \pmod{q}$ — this condition we call an *opportunity for action* and symbolize it by putting a directed edge $p \leftarrow q$ in the *opportunity graph*.

Not every group seizes every opportunity! The opportunities that are seized by a particular group we call its *action graph*, a subgraph of the opportunity graph. We explain the situation by discussing the cases p, pq, and pqr covered by Tables 16.2 and 16.3.

The case $p : q$ is exemplified by $11 : 5$, for which the relations are

$$1 = a^{11} = b^5,$$
$$b^{-1}ab = b^{3 \text{ or } 9 \text{ or } 5 \text{ or } 4},$$

since 3, $9 \equiv 3^2$, $5 \equiv 3^3$, and $4 \equiv 3^4$ are the numbers of order 5 (mod 11). However, it doesn't matter which one we choose, since we can pass from one to another by replacing b by b^2, b^3, or b^4. For similar reasons there is only one group named $p : q$ (which exists when $p \equiv 1$ (mod q)).

There is similarly only one group named $p : (q \times r)$ (existing when $p \equiv 1$ (mod qr)). An example is $31 : (5 \times 3)$, for which the relations are

$$1 = a^{31} = b^5 = c^3,$$
$$bc = cb,$$
$$b^{-1}ab = a^{2 \text{ or } 4 \text{ or } 8 \text{ or } 16},$$
$$c^{-1}ac = a^{5 \text{ or } 25},$$

since we can pass between $2, 4, 8$, and 16 by choice of b, and between 5 and 25 by choice of c. The same happens in any other "2 in" case. However, in the "2 out" and "3" cases, there are $r - 1$ groups of form $(p \times q) : r$, exemplified by $(31 \times 11) : 5_{i,j}$, for which the relations are

$$1 = a^{31} = b^{11} = c^5,$$
$$ab = ba,$$
$$c^{-1}ac = a^i,$$
$$c^{-1}bc = b^j.$$

In this case we can vary c to choose *either* between i in $\{2, 4, 8, 16\}$ *or* between j in $\{3, 9, 5, 4\}$, but not both, so there are $4 = r - 1$ distinct groups of this form.

For a general squarefree group the analysis is similar. Hölder shows that the total number of groups is given by the formula

$$\mathrm{gnu}(n) = \sum_{d \mid n} \prod_{p \mid d} \frac{(p^{\mathrm{opp}(p,e)} - 1)}{(p - 1)},$$

where $de = n$, $p \mid d$ and $\mathrm{opp}(p, e)$ is the number of opportunities for p to act on the primes dividing e.

Groups of Order pq^2

Hölder also enumerated the groups of order pq^2. He found that for a triprime (a number with three prime factors) of form $n = pq^2 > 12$, $\text{gnu}(n)$ is

$$
\begin{array}{ll}
2 & \text{if we have none of } p|q-1, p|q+1, q|p-1; \\
3 & \text{if both } p, q > 2 \text{ and } p|q+1; \\
4 & \text{if } p > 3, \ q|p-1, \text{ but not } q^2|p-1; \\
5 & \text{if } q^2|p-1; \text{ and finally} \\
\left[\frac{p+9}{2}\right] & \text{if } p|q-1,
\end{array}
$$

the corresponding groups being

$$
\begin{array}{l}
q \times q \times p, \ q^2 \times p \\
q \times q \times p, \ q^2 \times p, \ (q \times q) : p \\
q \times q \times p, \ q^2 \times p, \ (p:q) \times q, \ p : (q \times q) \\
q \times q \times p, q^2 \times p, (p:q) \times q, p : (q \times q), p : q^2 \\
q \times q \times p, \ q^2 \times p, \ q \times (q:p), \ q^2 : p, \ (q \times q) : p_{i,j}.
\end{array}
$$

In the last case, the automorphism is $a \mapsto a^i$, $b \mapsto b^j$ where i and j are numbers of order $p \pmod{q}$.

To help us count the groups of this last form, we study the example $(11 \times 11) : 5_{i,j}$, when i and j range over the set $\{3, 9, 5, 4\}$. Since i and j can be replaced by i^n and j^n, all that matters is the number m for which $j \equiv i^m \pmod{5}$. Since i and j can be interchanged, the cases m and m' are the same if $mm' \equiv 1 \pmod{5}$. For general p, m and m' are mutually inverse members of a cyclic group of order $p - 1$, which gives the integer part of $\frac{p+1}{2}$ for the number of groups of this form, and so the integer part of $\frac{p+9}{2}$ for the total number of groups in the last line.

The Group Number Function gnu(n)

We have already mentioned that the appendix to this chapter tabulates the number of abstract groups of each order $n < 2010$. In a forthcoming paper Conway, Dietrich, and O'Brien [5] study this group number function, gnu(n), in more detail. They note that the initial values approximately depend only on what we call the multiprimality of n (by which we mean its number of prime factors, counting repetitions), namely gnu $=$

1	for all prime numbers,
$2, 2, 2, 2, 2, 1, 2, 2, 2, 2, 1, 2, 1, 2, 2, ...$	for the first few biprimes,
$5, 5, 5, 5, 5, 4, 4, 6, 4, 2, 5, 5, 4, 4, 5, ...$	for triprimes,
$14, 15, 14, 15, 14, 14, 15, 13, 13, 15, 15, 12, 10, 16, ...$	for quadriprimes,
$51, 52, 50, 52, 45, 43, 47, 55, 57, 42, 37, 52, 51, 67, ...$	for quinqueprimes,
$267, 231, 197, 238, 177, 197, 208, ...$	for sextiprimes,

where, of course, by a k-prime we mean a number whose multiprimality is k. It seems that when the multiprimality of n is k, gnu(n) is approximately the kth Bell number, its "expected value." The expected values for multiprimality 1, 2, 3, 4, 5, 6, ... are 1, 2, 5, 15, 52, 203, ...

However, for larger values this rule is not always reliable. It's true that for biprimes gnu(n) is always 1 or 2 but it's usually 1, not 2, while for triprimes and above gnu(n) can be as small as 1 and can also be arbitrarily large.

The state of knowledge for prime powers is

$$\text{gnu}(p) = 1, \text{gnu}(p^2) = 2, \text{gnu}(p^3) = 5, \text{gnu}(p^4) = 15$$

(the expected values), except that gnu(16) $= 14$. For fifth, sixth, and seventh powers, we have exact formulae. Usually

$$\begin{aligned}
\text{gnu}(p^5) &= 61 + 2p + 2\gcd(p-1,3) + \gcd(p-1,4), \\
\text{gnu}(p^6) &= 3p^2 + 39p + 344 + 24\gcd(p-1,3) + 11\gcd(p-1,4) \\
&\quad +2\gcd(p-1,5), \\
\text{gnu}(p^7) &= 3p^5 + 12p^4 + 44p^3 + 170p^2 + 707p + 2455 \\
&\quad +(4p^2 + 44p + 291)\gcd(p-1,3) \\
&\quad +(p^2 + 19p + 135)\gcd(p-1,4) \\
&\quad +(3p+31)\gcd(p-1,5) + 4\gcd(p-1,7) \\
&\quad +5\gcd(p-1,8) + \gcd(p-1,9),
\end{aligned}$$

but the first few cases are exceptions:

$$\text{gnu}(2^5) = 51,$$
$$\text{gnu}(3^5) = 67,$$
$$\text{gnu}(2^6) = 267,$$
$$\text{gnu}(3^6) = 504,$$
$$\text{gnu}(2^7) = 2328,$$
$$\text{gnu}(3^7) = 9310,$$
$$\text{gnu}(5^7) = 34{,}297.$$

The only higher powers we know exactly are

$$\text{gnu}(256) = 56{,}092,$$
$$\text{gnu}(512) = 10{,}494{,}213,$$
$$\text{gnu}(1024) = 49{,}487{,}365{,}422,$$

but there is an asymptotic estimate

$$\text{gnu}(p^n) \cong p^{2n^3/27 + O(n^{5/2})},$$

originally by Higman [19] and Sims [25], improved by M.F. Newman and C. Seeley (private communication). Pyber [23] has deduced that

$$\text{gnu}(n) \le n^{(2/27 + o(1))\mu(n)^2},$$

where $\mu(n)$ is the largest exponent in the prime-power factorization of n.

There is a sense in which "almost all groups have order a power of 2." For example, there are 423,164,719 groups of orders < 2048 other than 1024 and 49,487,365,422 of order 1024. If a group is selected at random from all the groups of order less than 2048, the odds are more than 100-to-1 that it will have order 1024. The first n for which $\text{gnu}(n)$ is unknown is 2048, but it is known that $\text{gnu}(2048)$ strictly exceeds 1,774,274,116,992,170, which is the exact number of groups of order 2048 that have "exponent 2 nilpotency class 2."

The gnu-Hunting Conjecture: Hunting moas

The *gnu-hunting conjecture* is that every positive integer n arises as $\text{gnu}(m)$ for some m. R. Keith Dennis has verified that every $n < 10,000,000$ is in fact of the form $\text{gnu}(m)$ for some square-free m. Conway, Dietrich, and O'Brien define $\text{moa}(n)$ (the minimal order attaining n) to be the least m with $\text{gnu}(m) = n$ and have hunted down the following moas:

	+0	+1	+2	+3	+4	+5	+6	+7	+8	+9
0		1	4	75	28	8	42	375	510	308
10	90	140	88	56	16	24	100	675	156	1029
20	820	1875	6321	294	546	2450	2550	1210	2156	1380
30	270	?	630	?	450	616	612	180	1372	264
40	280	420	176	112	392	108	252	120	2730	300
50	72	32	48	656	272	162	500	168	4650	6875
60	378	312	702	3630	1596	?	588	243	882	1215
70	4100	3660	1638	?	?	2420	2964	1092	?	3612
80	6050	6820	?	?	2394	?	?	?	2028	4140
90	?	?	?	?	?	?	6930	6498	4950	1188
100	3822									

The best guess for the first unknown entry $\text{moa}(31)$ is $11,774$. They also make the *galloping gnu conjecture*, stating that iterating the gnu function always leads to 1 in finitely many steps. For $n < 2048$ at most five steps are needed, the hardest case to verify being

$$1024 \to 49,487,365,422 \to 240 \to 208 \to 51 \to 1.$$

However, it seems that the galloping gnu conjecture will be extremely difficult to prove.

Appendix: The Number of Groups to Order 2009

The table on the next four pages, largely taken from a paper by Besche, Eick, and O'Brien [1], lists the number of groups for each order below 2010. There are many interesting patterns in the numbers, but exact formulae are known only for the few cases given in the text. The table has been extended privately to 2047, but finding the number for 2048 will be very difficult.

	+0	+1	+2	+3	+4	+5	+6	+7	+8	+9
0		1	1	1	2	1	2	1	5	2
10	2	1	5	1	2	1	14	1	5	1
20	5	2	2	1	15	2	2	5	4	1
30	4	1	51	1	2	1	14	1	2	2
40	14	1	6	1	4	2	2	1	52	2
50	5	1	5	1	15	2	13	2	2	1
60	13	1	2	4	267	1	4	1	5	1
70	4	1	50	1	2	3	4	1	6	1
80	52	15	2	1	15	1	2	1	12	1
90	10	1	4	2	2	1	231	1	5	2
100	16	1	4	1	14	2	2	1	45	1
110	6	2	43	1	6	1	5	4	2	1
120	47	2	2	1	4	5	16	1	2328	2
130	4	1	10	1	2	5	15	1	4	1
140	11	1	2	1	197	1	2	6	5	1
150	13	1	12	2	4	2	18	1	2	1
160	238	1	55	1	5	2	2	1	57	2
170	4	5	4	1	4	2	42	1	2	1
180	37	1	4	2	12	1	6	1	4	13
190	4	1	1543	1	2	2	12	1	10	1
200	52	2	2	2	12	2	2	2	51	1
210	12	1	5	1	2	1	177	1	2	2
220	15	1	6	1	197	6	2	1	15	1
230	4	2	14	1	16	1	4	2	4	1
240	208	1	5	67	5	2	4	1	12	1
250	15	1	46	2	2	1	56092	1	6	1
260	15	2	2	1	39	1	4	1	4	1
270	30	1	54	5	2	4	10	1	2	4
280	40	1	4	1	4	2	4	1	1045	2
290	4	2	5	1	23	1	14	5	2	1
300	49	2	2	1	42	2	10	1	9	2
310	6	1	61	1	2	4	4	1	4	1
320	1640	1	4	1	176	2	2	2	15	1
330	12	1	4	5	2	1	228	1	5	1
340	15	1	18	5	12	1	2	1	12	1
350	10	14	195	1	4	2	5	2	2	1
360	162	2	2	3	11	1	6	1	42	2
370	4	1	15	1	4	7	12	1	60	1
380	11	2	2	1	20169	2	2	4	5	1
390	12	1	44	1	2	1	30	1	2	5
400	221	1	6	1	5	16	6	1	46	1
410	6	1	4	1	10	1	235	2	4	1
420	41	1	2	2	14	2	4	1	4	2
430	4	1	775	1	4	1	5	1	6	1
440	51	13	4	1	18	1	2	1	1396	1
450	34	1	5	2	2	1	54	1	2	5
460	11	1	12	1	51	4	2	1	55	1
470	4	2	12	1	6	2	11	2	2	1
480	1213	1	2	2	12	1	261	1	14	2
490	10	1	12	1	4	4	42	2	4	1
500	56	1	2	1	202	2	6	6	4	1

	+0	+1	+2	+3	+4	+5	+6	+7	+8	+9
510	8	1	10494213	15	2	1	15	1	4	1
520	49	1	10	1	4	6	2	1	170	2
530	4	2	9	1	4	1	12	1	2	2
540	119	1	2	2	246	1	24	1	5	4
550	16	1	39	1	2	2	4	1	16	1
560	180	1	2	1	10	1	2	49	12	1
570	12	1	11	1	4	2	8681	1	5	2
580	15	1	6	1	15	4	2	1	66	1
590	4	1	51	1	30	1	5	2	4	1
600	205	1	6	4	4	7	4	1	195	3
610	6	1	36	1	2	2	35	1	6	1
620	15	5	2	1	260	15	2	2	5	1
630	32	1	12	2	2	1	12	2	4	2
640	21541	1	4	1	9	2	4	1	757	1
650	10	5	4	1	6	2	53	5	4	1
660	40	1	2	2	12	1	18	1	4	2
670	4	1	1280	1	2	17	16	1	4	1
680	53	1	4	1	51	1	15	2	42	2
690	8	1	5	4	2	1	44	1	2	1
700	36	1	62	1	1387	1	2	1	10	1
710	6	4	15	1	12	2	4	1	2	1
720	840	1	5	2	5	2	13	1	40	504
730	4	1	18	1	2	6	195	2	10	1
740	15	5	4	1	54	1	2	2	11	1
750	39	1	42	1	4	2	189	1	2	2
760	39	1	6	1	4	2	2	1	1090235	1
770	12	1	5	1	16	4	15	5	2	1
780	53	1	4	5	172	1	4	1	5	1
790	4	2	137	1	2	1	4	1	24	1
800	1211	2	2	1	15	1	4	1	14	1
810	113	1	16	2	4	1	205	1	2	11
820	20	1	4	1	12	5	4	1	30	1
830	4	2	1630	2	6	1	9	13	2	1
840	186	2	2	1	4	2	10	2	51	2
850	10	1	10	1	4	5	12	1	12	1
860	11	2	2	1	4725	1	2	3	9	1
870	8	1	14	4	4	5	18	1	2	1
880	221	1	68	1	15	1	2	1	61	2
890	4	15	4	1	4	1	19349	2	2	1
900	150	1	4	7	15	2	6	1	4	2
910	8	1	222	1	2	4	5	1	30	1
920	39	2	2	1	34	2	2	4	235	1
930	18	2	5	1	2	2	222	1	4	2
940	11	1	6	1	42	13	4	1	15	1
950	10	1	42	1	10	2	4	1	2	1
960	11394	2	4	2	5	1	12	1	42	2
970	4	1	900	1	2	6	51	1	6	2
980	34	5	2	1	46	1	4	2	11	1
990	30	1	196	2	6	1	10	1	2	15
1000	199	1	4	1	4	2	2	1	954	1

	+0	+1	+2	+3	+4	+5	+6	+7	+8	+9
1010	6	2	13	1	23	2	12	2	2	1
1020	37	1	4	2	49487365422	4	66	2	5	19
1030	4	1	54	1	4	2	11	1	4	1
1040	231	1	2	1	36	2	2	2	12	1
1050	40	1	4	51	4	2	1028	1	5	1
1060	15	1	10	1	35	2	4	1	12	1
1070	4	4	42	1	4	2	5	1	10	1
1080	583	2	2	6	4	2	6	1	1681	6
1090	4	1	77	1	2	2	15	1	16	1
1100	51	2	4	1	170	1	4	5	5	1
1110	12	1	12	2	2	1	46	1	4	2
1120	1092	1	8	1	5	14	2	2	39	1
1130	4	2	4	1	254	1	42	2	2	1
1140	41	1	2	5	39	1	4	1	11	1
1150	10	1	157877	1	2	4	16	1	6	1
1160	49	13	4	1	18	1	4	1	53	1
1170	32	1	5	1	2	2	279	1	4	2
1180	11	1	4	3	235	2	2	1	99	1
1190	8	2	14	1	6	1	11	14	2	1
1200	1040	1	2	1	13	2	16	1	12	5
1210	27	1	12	1	2	69	1387	1	16	1
1220	20	2	4	1	164	4	2	2	4	1
1230	12	1	153	2	2	1	15	1	2	2
1240	51	1	30	1	4	1	4	1	1460	1
1250	55	4	5	1	12	2	14	1	4	1
1260	131	1	2	2	42	3	6	1	5	5
1270	4	1	44	1	10	3	11	1	10	1
1280	1116461	5	2	1	10	1	2	4	35	1
1290	12	1	11	1	2	1	3609	1	4	2
1300	50	1	24	1	12	2	2	1	18	1
1310	6	2	244	1	18	1	9	2	2	1
1320	181	1	2	51	4	2	12	1	42	1
1330	8	5	61	1	4	1	12	1	6	1
1340	11	2	4	1	11720	1	2	1	5	1
1350	112	1	52	1	2	2	12	1	4	4
1360	245	1	4	1	9	5	2	1	211	2
1370	4	2	38	1	6	15	195	15	6	2
1380	29	1	2	1	14	1	32	1	4	2
1390	4	1	198	1	4	8	5	1	4	1
1400	153	1	2	1	227	2	4	5	19324	1
1410	8	1	5	4	4	1	39	1	2	2
1420	15	4	16	1	53	6	4	1	40	1
1430	12	5	12	1	4	2	4	1	2	1
1440	5958	1	4	5	12	2	6	1	14	4
1450	10	1	40	1	2	2	179	1	1798	1
1460	15	2	4	1	61	1	2	5	4	1
1470	46	1	1387	1	6	2	36	2	2	1
1480	49	1	24	1	11	10	2	1	222	1
1490	4	3	5	1	10	1	41	2	4	1
1500	174	1	2	2	195	2	4	1	15	1

	+0	+1	+2	+3	+4	+5	+6	+7	+8	+9
1510	6	1	889	1	2	2	4	1	12	2
1520	178	13	2	1	15	4	4	1	12	1
1530	20	1	4	5	4	1	408641062	1	2	60
1540	36	1	4	1	15	2	2	1	46	1
1550	16	1	54	1	24	2	5	2	4	1
1560	221	1	4	1	11	1	30	1	928	2
1570	4	1	10	2	2	13	14	1	4	1
1580	11	2	6	1	697	1	4	3	5	1
1590	8	1	12	5	2	2	64	1	4	2
1600	10281	1	10	1	5	1	4	1	54	1
1610	8	2	11	1	4	1	51	6	2	1
1620	477	1	2	2	56	5	6	1	11	5
1630	4	1	1213	1	4	2	5	1	72	1
1640	68	2	2	1	12	1	2	13	42	1
1650	38	1	9	2	2	2	137	1	2	5
1660	11	1	6	1	21507	5	10	1	15	1
1670	4	1	34	2	60	2	4	5	2	1
1680	1005	2	5	2	5	1	4	1	12	1
1690	10	1	30	1	10	1	235	1	6	1
1700	50	309	4	2	39	7	2	1	11	1
1710	36	2	42	2	2	5	40	1	2	2
1720	39	1	12	1	4	3	2	1	47937	1
1730	4	2	5	1	13	1	35	4	4	1
1740	37	1	4	2	51	1	16	1	9	1
1750	30	2	64	1	2	14	4	1	4	1
1760	1285	1	2	1	228	1	2	5	53	1
1770	8	2	4	2	2	4	260	1	6	1
1780	15	1	110	1	12	2	4	1	12	1
1790	4	5	1083553	1	12	1	5	1	4	1
1800	749	1	4	2	11	3	30	1	54	13
1810	6	1	15	2	2	9	12	1	10	1
1820	35	2	2	1	1264	2	4	6	5	1
1830	18	1	14	2	4	1	117	1	2	2
1840	178	1	6	1	5	4	4	1	162	2
1850	10	1	4	1	16	1	1630	2	2	2
1860	56	1	10	15	15	1	4	1	4	2
1870	12	1	1096	1	2	21	9	1	6	1
1880	39	5	2	1	18	1	4	2	195	1
1890	120	1	9	2	2	1	54	1	4	4
1900	36	1	4	1	186	2	2	1	36	1
1910	6	15	12	1	8	1	4	5	4	1
1920	241004	1	5	1	15	4	10	1	15	2
1930	4	1	34	1	2	4	167	1	12	1
1940	15	1	2	1	3973	1	4	1	4	1
1950	40	1	235	11	2	1	15	1	6	1
1960	144	1	18	1	4	2	2	2	203	1
1970	4	15	15	1	12	2	39	1	4	1
1980	120	1	2	2	1388	1	6	1	13	4
1990	4	1	39	1	2	5	4	1	66	1
2000	963	1	8	1	10	2	4	4	12	2

Part III

Repeating Patterns in Other Spaces

Introduction to Part III

Part I of our book is accessible to those interested in mathematics but with no particular mathematical skills, while we expect readers of Part II to know some group theory (a first course suffices for all but a few tricky details). We expect that Part III will be completely understood only by a few professional mathematicians. Once again, however, much of it can be read with profit by some readers, and many more will enjoy inspecting our pretty pictures.

In Chapter 2 we listed the 17 plane crystallographic groups. Chapters 22–25 perform the same service for their three-dimensional analogs, the 219 crystallographic space groups. The most interesting of these are the 35 groups that we call "prime," which are discussed in Chapters 22 and 23.

The orbifolds of those space groups in which only the identity element fixes a point are the ten platycosms, or "flat universes," which are the three-dimensional analogs of the torus and Klein bottle. They are illustrated in Chapter 24, which derives them from the complete list of 184 composite space groups in Chapter 25.

Finally, Chapter 26 discusses dimensions higher than three. The enumeration of the finite groups in four dimensions is due to several authors, starting with Goursat in 1889. The regular polytopes in high dimensions were listed by Schäfli in 1852, while Coxeter enumerated groups generated by reflections in all dimensions. Our book is the first to describe the complete list of four-dimensional Archimedean polytopes found by Conway and Guy in 1965. The book ends with a few remarks on very high dimensions indeed!

– 17 –

Introducing Hyperbolic Groups

In Chapter 3, we saw that Euclidean plane groups are classified by signatures that cost exactly $2, and in Chapter 4 we saw that the signatures costing less than $2 perform the same service for the spherical groups.

What about signatures like ∗**732** that cost more than $2 or, equivalently, have negative Euler characteristic and *ch*? The answer is that they classify the corresponding symmetry groups in the geometry of the non-Euclidean plane discovered by Lobachevski in 1829, in which there are many lines through a point that don't meet a given line. Nowadays, mathematicians use the name "hyperbolic plane" introduced by Cayley. But you won't need to know any hyperbolic geometry, because we can say it all in pictures!

In fact, there has already been some non-Euclidean geometry in our book—that of the sphere, in which every "straight line" (i.e., great circle) meets every other. The fact that the hyperbolic plane is not Euclidean won't stop us from drawing pictures, just as nothing prevents us from drawing pictures of a sphere on our Euclidean pages.

Just as geographers "project" the spherical surface of the earth on their flat atlas pages, we can map the hyperbolic plane onto the Euclidean.

No Projection Is Perfect!

As you probably know, Mercator's projection distorts areas. Everybody knows that it makes Greenland look larger than the whole of

Parallel Principle

In the Euclidean plane, there is just one line parallel to a given line through a given point. In spherical geometry there are no parallels, as shown above. In hyperbolic geometry, there are many lines through a point that don't meet a given line.

(opposite page) A hyperbolic pattern of type **23**×.

219

South America, but even worse, it makes Antarctica infinite! Its great advantage is that it preserves angles and directions, which is what made it a boon to the navigators for whom it was designed. On the other hand, the cylindrical projection discovered by Archimedes (though called "Lambert's projection" by cartographers), preserves areas but grossly deforms angles, directions, and shapes. If you find a projection that preserves one thing important to you, you can be sure that it will disturb another! There is no perfect projection from a spherical surface to the Euclidean plane.

The orthogonal projection distorts areas and angles and only shows half of the earth at a time!

The Mercator projection preserves angles but greatly distorts areas!

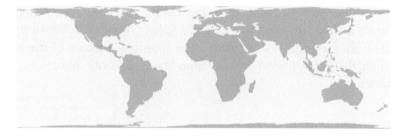

The less common cylindrical projection preserves areas but greatly distorts horizontal distances near the poles.

The two most common 'projections' of the hyperbolic plane take it to a disk in the Euclidean plane. One of them, due to Beltrami and Klein, takes hyperbolic straight lines to (segments of) Euclidean

Figure 17.1. The Beltrami-Klein disk model of the hyperbolic plane, which is similar to the orthogonal projection of the sphere.

ones. (See Figure 17.1.) This is really the most natural projection, because it is the way the hyperbolic plane would appear if you viewed it from a point in hyperbolic 3-space.

However, the other method, usually credited to Poincaré, is more widely known to both mathematicians and artists. It is the projection used, for example, in Maurits C. Escher's famous "Circle Limit" engravings. (See Figure 17.4.) It preserves angles but takes hyperbolic straight lines to arcs of circles perpendicular to the boundary. (See Figure 17.2.) The reason this less-natural projection is so often used is that it shows more of the plane. It is equivalent by inversion to the upper half-plane model used by mathematicians—shown in Figure 17.3—which also preserves angles.

Figure 17.2. The Poincaré disk model of the hyperbolic plane, which preserves angles.

Figure 17.3. The upper half-plane model, which also preserves angles.

Poincaré and Klein

Here are drawings of the same tiling in the hyperbolic plane, shown in different models:

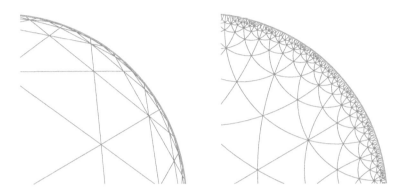

In the Klein model (left) straight lines on the hyperbolic plane are shown as straight lines in the drawing. The drawing appears to be in perspective and shows how the hyperbolic plane would really look were we to view it from within hyperbolic 3-space.

The Poincaré model (right) and the upper half-plane model (below) preserve angles; straight lines appear to be circular arcs perpendicular to the boundary of the drawing. We can see much further than we could in the Klein model.

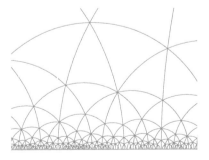

Figure 17.4. At first glance, M. C. Escher's *Circle Limit IV*, often called "Angels and Devils," seems to have signature **4∗3**, but in fact it has signature **∗3333**.

Analyzing Hyperbolic Patterns

It is a remarkable fact that essentially all the orbifold theory continues to work in the hyperbolic case. This even includes the proofs, so we can use it immediately to analyse the symmetries of the "Circle Limit" engravings. The one that personally intrigues us most is *Circle Limit IV*, often called "Angels and Devils" (Figure 17.4). Remembering our advice from Chapter 3, you should first concentrate on the mirrors; you will find that they cut the (hyperbolic) plane into quadrilaterals with four angles of $\frac{\pi}{3}$. If there were no further symmetries, the answer would be the kaleidoscopic group $*\mathbf{3333}$. However, Escher cleverly chose this particular quadrilateral so that he could suggest the 4-fold rotational symmetry that extends this group to $\mathbf{4}*\mathbf{3}$. Look at the way that the four devils' wingtips alternate with four angels' ones at the center of each of these quadrilaterals, and it will be easy for you to conclude that its symmetry group is $\mathbf{4}*\mathbf{3}$. The total cost of this is

$$^{\$}\frac{3}{4} + 1 + \frac{1}{3} = 2\frac{1}{12}, \text{ making } ch = ^{\$}-\frac{1}{12},$$

exemplifying our assertion that the signatures for hyperbolic groups are those with negative orbifold Euler characteristics.

However, Escher is more devilish than you think. His angels and devils *don't* in fact have symmetry group $\mathbf{4}*\mathbf{3}$, but only its kaleidoscopic subgroup $*\mathbf{3333}$. And we are no angels either; although we told no lies, we must admit that we deliberately tried to deceive you. Did we succeed?

The subtle thing that Escher did in this wonderful work was to make every fourth figure—either angel or devil—face away from the viewer. This does not easily show itself in poor reproductions, but it is well worth observing. If you look closely at one of those quadrilaterals, you will find that one of its four devils seems to have no eyes (this is because you can't see through the back of his head!), as does one of its four angels. You are also looking at the backs of their garments, although this might be hard to see.

Now we evaluate the true cost and characteristic of *Circle Limit IV*:

$$\text{cost} = {}^{\$}1 + \frac{1}{3} + \frac{1}{3} + \frac{1}{3} + \frac{1}{3} = 2\frac{1}{3}, ch = {}^{\$}-\frac{1}{3}.$$

What Do Negative Characteristics Mean?

In Chapter 6 we saw that the characteristic of a pattern that has a finite number g of symmetries was $\frac{2}{g}$, and we had already seen that the characteristic of an infinite Euclidean pattern was 0, which can be regarded as the case $g = \infty$ of this formula, which therefore works for all positive or zero characteristics. What do negative characteristics tell us about the number of symmetries of hyperbolic patterns like Escher's *Circle Limit IV*? Since the correct characteristic for this pattern is $-\frac{1}{3} = \frac{2}{-6}$, it seems to be trying to tell us that the number of symmetries is -6, which can hardly be! If we hadn't noticed the subtlety, we would have found the equally nonsensical answer -24. Let's be honest: the order formula fails for hyperbolic groups. But it's the only thing that does. This one failure does not matter too much because we know how many symmetries there are: all hyperbolic patterns have infinitely many.

Even the order formula gets something right. Look at the two "wrong" answers, -6 and -24, for *Circle Limit IV* with and without its subtlety. The second is four times the first, and it is indeed true that a person who doesn't notice the subtlety will think the pattern is four times as symmetric as it really is.

This corresponds to the group-theoretical fact that ∗**3333** has index 4 in **4**∗**3**. Our "incorrect" order formula always gets the correct indices. Why is the formula wrong, and why does it get indices right?

The reason that indices continue to work is that if a group **H** has index n in a larger group **G**, then **H**'s orbifold is n times as big as **G**'s (in the precise sense that there is an n-to-1 map from the former to the later), and so it is obvious that **H**'s characteristic is n times as big as **G**'s. The way we proved the order formula for **G** was just to apply this fact with **H** = 1. We cannot do this in the hyperbolic case, since there the trivial group has infinite index (and even if we could, we would get the meaningless answer ∞/∞).

With this one exception, all the ideas and proofs we've used continue to work in the hyperbolic cases, and they need no further justification.

Types of Coloring, Tiling, and Group Presentations

Almost everything that we did earlier in the book for spherical, Euclidean, and frieze types extends easily to the hyperbolic case, with the exception that there are now infinitely many starting groups. For example, presentations can be found in exactly the same way and their generators indicated by annotating the signature as usual.

Here is, for example, a Cayley graph of the group with type **732**; it looks almost exactly like the similar graph for **532**—the graphs *are* the same on the orbifold.

A Cayley graph of the hyperbolic group **732**.

By replacing the names of generators by permutations that satisfy the relations, we obtain an annotated symbol that always characterizes the color type. In simple cases our notations **G/H/K**, **G/K**, and **G//K** suffice. Duality groups can be handled in the same way.

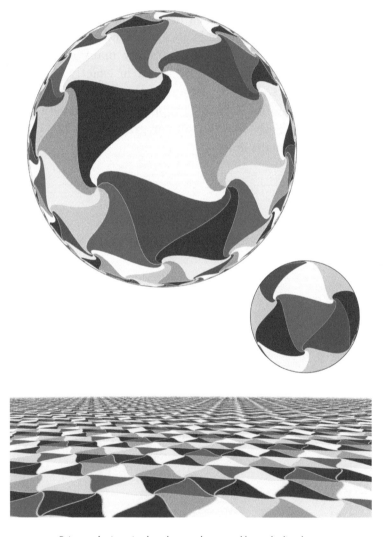

Prime colorings in the plane, sphere, and hyperbolic plane.

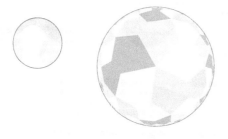

A 3-coloring of the spherical group **3∗2**; a 4-coloring of the Euclidean group **4∗2**; a 5-coloring of the hyperbolic group **5∗2**.

Our method of classifying Heesch and isohedral tilings of a given symmetry type by cutting open the corresponding orbifold is still valid. Since, in fact, this only depends on the topology of the orbifold, we've already classified all Heesch and isohedral tilings with the hyperbolic symmetries of the forms **pqr**, **pqrs**, **pq∗**, **pq×**, **p∗q**, **p∗qr**, **∗pqr**, and **∗pqrs**!

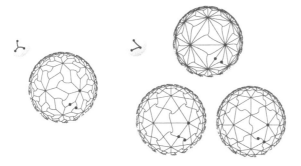

The Heesch tilings with symmetry **732**.

Where Are We?

In this chapter, we introduced the hyperbolic groups and showed the ways they are the same as the Euclidean and spherical groups. In the next chapter, we'll discuss issues that are particularly hyperbolic.

– 18 –

More on Hyperbolic Groups

In this chapter, we consider a number of topics that are special to the hyperbolic groups.

Which Signatures Are Really the Same?

The symbols $*\mathbf{632}$, $*\mathbf{623}$, $*\mathbf{362}$, $*\mathbf{326}$, and $*\mathbf{263}$, $*\mathbf{236}$ all mean the same thing. What's the general rule for telling when two signatures are "really the same"? For the reasons explained in Chapter 3, "the same" should mean that the corresponding groups are isotopic. The answer is best expressed by describing various operations that don't change the isotopy type. Namely, the group represented by the typical signature

$$\ldots \circ\circ\circ\mathbf{ABC} \ldots \; *\mathbf{a_1 b_1 c_1}\ldots*\mathbf{a_2 b_2 c_2}\ldots * \mathbf{a_n b_n c_n}\ldots \times \times$$

will be unchanged up to isotopy if we

- exchange an $\circ$ and an $\times$ for three $\times$'s,

- freely permute the digits $\mathbf{A,B,C,\ldots}$ that correspond to gyrations,

- cyclically permute the digits $\mathbf{a_k, b_k, c_k}, \ldots$ in any one kaleidoscope,

- freely permute the portions $*\mathbf{a_1 b_1 c_1} \ldots, *\mathbf{a_2 b_2 c_2} \ldots, \ldots,$ $*\mathbf{a_n b_n c_n}\ldots$ of the signature corresponding to the individual kaleidoscopes,

---

(opposite page) For your viewing pleasure, this figure has signature $\times\times\times$. The precise symmetry is difficult to discern without examining the guide lines shown.

- simultaneously reverse the cyclic orders in all **n** kaleidoscopic portions.

If the orbifold surface is orientable (i.e., there is no × in the signature), this is all. But, we may also

- independently reverse the cyclic orders in individual kaleidoscopic portions, provided that the signature contains an ×.

This is all—if two signatures represent isotopic groups, then you can get from one to the other using only these operations. This is immediate from Thurston's way of constructing orbifolds, which we'll briefly describe later in this chapter. In particular, it implies the strong theorem that, if two orbifolds are homeomorphic in a way that preserves the orders of their cone points and kaleidoscopic points, they are actually isotopic.

Inequivalence and Equivalence Theorems

As we saw in Chapter 4, there are many isomorphisms between distinct spherical groups, for instance $* \cong \mathbf{22} \cong \times$, but in 1954 W. Nowacki proved that the 17 Euclidean groups are abstractly distinct using a number of ad hoc invariants [22]. There are, however, some isomorphisms between the frieze groups:

$$\mathbf{2}*\infty \cong \mathbf{22}\infty \cong *\infty\infty \cong \mathbf{D}_\infty, \infty\times \cong \infty\infty \cong \mathbf{C}_\infty.$$

(The remaining two groups have different structures: $*\mathbf{22}\infty \cong \mathbf{2} \times \mathbf{D}[\infty]$, $\infty\times \cong \mathbf{2} \times \mathbf{C}[\infty]$.) So such isomorphisms can happen between finite groups or groups with ∞ in their name, (i.e., those whose orbifolds are non-compact). We shall soon describe all these equivalences.

Groups with Compact Orbifolds

On the other hand, Thurston has given a uniform proof that infinite groups with compact orbifolds can only be isomorphic if they are isotopic. Thurston's proof is unpublished and quite complicated, since it involves his subtle geometrization techniques. We give a simple (but long!) group-theoretical proof in the appendix to this chapter.

Groups with Noncompact Orbifolds

Signatures also work for some hyperbolic groups for which the orbifold is not compact. It is usual to describe this class as those whose orbifolds "have finite volume," but we prefer to say "are almost-compact," since this also applies to their Euclidean analogs, the frieze groups (whose orbifolds have infinite area). (The almost-compact groups are those that can be compactified by adjoining the images in the orbifold of finitely many ideal points.)

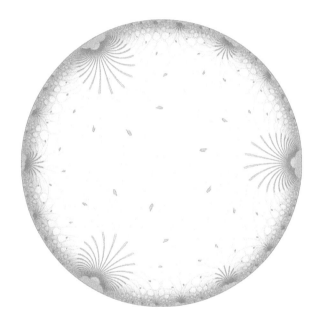

Figure 18.1. A pattern with non-compact orbifold and signature ∞**32**, isomorphic to $\mathbf{SL}_2(\mathbb{Z})$.

The signatures for groups that are almost compact but not compact are precisely those for which at least one digit is ∞. Figures 18.1 and 18.2 show the two best known examples, ∞**32** and ∗∞**32**, which are isomorphic to the modular group $\mathbf{SL}_2(\mathbb{Z})$ and the extended modular group $\mathbf{GL}_2(\mathbb{Z})$, respectively.

Geometrically, $\mathbf{GL}_2(\mathbb{Z})$ is the symmetry group of the packing of Ford circles in the upper half-plane model—Figure 18.3 shows this, as well as the Poincaré disk model. In the hyperbolic plane, these circles have infinite radius and are called *horocycles*.

Figure 18.2. A pattern with signature $*\infty32$, isomorphic to $\mathbf{GL}_2(\mathbb{Z})$.

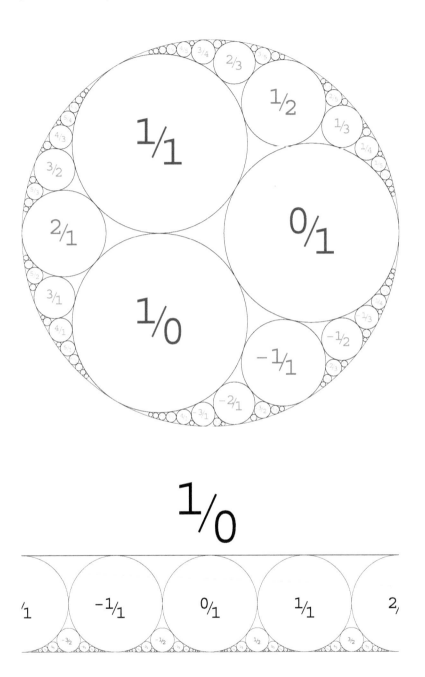

Figure 18.3. The Ford circles also have symmetry group $\mathbf{GL}_2(\mathbb{Z})$.

Abstract Isomorphisms between Almost-Compact Groups

Of course, we can only hope to find nontrivial isomorphisms between groups whose signatures both contain ∞. It is easy to find all such isomorphisms, because when ∞ is involved, the groups break up into free products. We first "normalize" so that the name of each kaleidoscope that contains an ∞ begins ∗∞... and call such a star an *infinite red star*. Then, the abstract group is not affected when we

- replace a wonder-ring (○) by two infinite gyration points (∞∞),

- replace a cross (×) by one infinite gyration point (∞),

- replace a simply-infinite kaleidoscope (∗∞) by two gyration points (**2**∞) of degrees 2 and ∞,

- trade each infinite red star after the first for a blue ∞,

- fuse an adjacent pair of red ∞'s and adjoin a blue **2**.

For instance, the groups ∗**ab** . . . ∞∗**cd** . . . ∞ and ∞∗**ab** . . . ∞**cd** . . . ∞ are abstractly isomorphic (here we have traded the second infinite red star for a blue ∞).

After all such replacements, the signature consists of numbers that indicate gyration points (some of which may be ∞) and kaleidoscopes (at most one of which includes ∞). If there is a kaleidoscope with more than one ∞, say

$$∗∞\mathbf{ab}...\mathbf{c}∞\mathbf{de}...\mathbf{f}∞\mathbf{gh}...\mathbf{i}∞ \; ... \; ... \; ∞\mathbf{xy}...\mathbf{z}$$

(with **a**, **b**, ..., **z** all finite), then we can also

- bodily permute the "finite blocks" **ab...c**, . . . , **xy...z** and/or

- reverse the digits in any of them.

For example, the above kaleidoscope can be replaced by

$$∗∞\mathbf{f}...\mathbf{ed}∞\mathbf{xy}...\mathbf{z}∞\mathbf{ab}...\mathbf{c}∞ \; ... \; ... \; ∞\mathbf{i}...\mathbf{hg}.$$

We omit the easy proof that these alterations account for all iso-morphisms. The isomorphisms arise because the presence of an ∞ causes the group to decompose as the free product of the separate groups defined by the local relations, but with one free generator deleted.

Existence and Construction

Recall that in the spherical case there were "bad" signatures like $*\mathbf{MN}$ and $\mathbf{MN}$ $(M \neq N)$ for which there were no corresponding groups. Are there any such "bad" signatures in the hyperbolic plane?

No! Thurston has shown that every signature with negative characteristic really does correspond to a group and moreover that if two groups have the same signature then one can be isotopically deformed until it is isometric to the other. He does this by cutting the putative orbifold into parts along geodesics (in increasing order of length) until each part is a "generalized triangle." His argument doesn't work for the last four spherical types ($*\mathbf{NN}$, $\mathbf{N}*$, $\mathbf{N}\times$, $\mathbf{NN}$) and the last four Euclidean ones ($**$, $*\times$, $\times\times$, $\circ$) since for them the decomposition involves digons rather than triangles. This is why it doesn't prove that orbifolds exist in the above bad cases.

The real angles of these "triangles" are prescribed numbers of the form $A\pi/n$. The imaginary ones are really line-segments at which one triangle should be attached to another, so the lengths of these segments should be equal in pairs. These are conditions on the "trigonometry" of these triangles, so the construction of all orbifolds satisfying them reduces to finding all solutions of a certain set of equations. Thurston shows that the space of solutions, counted up to scale, is homeomorphic to a Euclidean space $\mathbb{R}^n$ (and in particular is not empty). He also gives an elegant rule for the dimension n: namely, $\chi_3 = -n/3$, where χ_3 is the orbifold Euler characteristic of the orbifold obtained from the given one by replacing each of its numerals by 3. How many parameters has

$$4*\mathbf{235}*\mathbf{74}\times\times?$$

The answer is 13, since the characteristic of $\mathbf{3}*\mathbf{333}*\mathbf{33}\times\times$ is

$$2 - \left(\frac{2}{3} + 1 + \frac{1}{3} + \frac{1}{3} + \frac{1}{3} + 1 + \frac{1}{3} + \frac{1}{3} + 1 + 1\right) = -\frac{13}{3}.$$

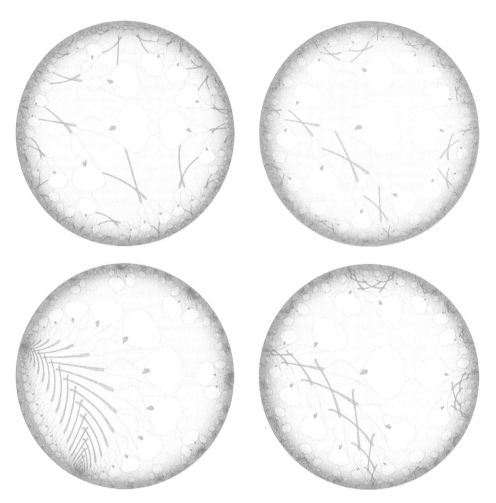

Parameters in the hyperbolic groups. The hyperbolic groups typically have a great deal of flexibility in their particular geometric realizations. For example, the group *3* has characteristic −1/3 and thus has one degree of freedom in its construction. In these images, we see how the spacing of the lilies can be varied.

The argument doesn't work for the last four spherical cases, *NN, N*, N×, and NN (where it would make the number of parameters negative!), and the last four Euclidean ones, **, *×, ××, and ○, which have respectively one, one, one, and two parameters rather than the zero given by the rule.

Enumerating Hyperbolic Groups

It is now easy to list all the possibilities for groups of any given characteristic, except that in the hyperbolic case the number of answers can be arbitrarily large. However, it is always finite, which proves that the hyperbolic groups can be well-ordered so that their characteristics are monotonic. The higher-dimensional analogs of this result have also been proved but are more subtle.

Table 18.1 lists all possibilities that have characteristic at least $-\frac{1}{8}$. We use ordinal numbers, which may be infinite, to enumerate

Number	$-1/char$	Groups
$\omega+1$	84	*237
$\omega+2$	48	*238
$\omega+3$	42	237
$\omega+4$	40	*245
$\omega+8$	24	*2 3 12, *246, *334, 3*4, 238
$\omega+11$	20	*2 3 15, *255, 5*2, 245
$\omega+$	18 2/3	*247
$\omega+15$	18	*2 3 18, 239
$\omega+21$	16	*2 3 24, *248
$\omega+27$	15	*2 3 30, *256, *335, 3*5, 2 3 10
$\omega+33$	14 2/5	*2 3 36, *249
$\omega+57$	13 1/3	*2 3 60, *2 4 10
$\omega+63$	13 1/5	*2 3 66, 2 3 11
$\omega+102$	12 8/11	*2 3 105, *257
$\omega+129$	12 4/7	*2 3 132, *2 4 11
...	...	...
$\omega\cdot 2$	12	*23∞, *2 4 12, *266, 6*2, *336, 3*6, *344, 4*3, *2223, 2*23, 2 3 12, 246, 334
$\omega\cdot 2+$	9 3/5	*2 4 24, *268, *338, 3*8, 2 3 16
finite	9 13/23	*2 5 11
	9 1/3	*2 4 28, *277, 7*2, 247
	9 3/11	2 3 17
	9 3/13	*2 4 30, *2 5 12, *345
	9	*2 4 36, *269, *339, 3*9, 2 3 18
	8 28/29	*2 5 13
	8 3/4	*2 5 14
	8 10/13	2 3 19
	8 8/13	*2 4 56, *278
	8 4/7	*2 4 60, *2 5 15, *2 6 10, *3 3 10, 3*10, 2 3 20

Table 18.1. The first few hyperbolic groups.

Number	$-1/char$	Groups
$\omega \cdot 2+$	8 8/19	*2 4 80, *2 5 16
finite	8 2/5	2 3 21
	8 12/41	*2 5 17
	8 1/4	*2 4 132, *2 6 11, *3 3 11, 3*11, 2 3 22
	8 2/11	*2 4 180, *2 5 18
	8 4/31	*2 4 252, *279
	8 2/17	2 3 23
	8 4/47	*2 4 380, *2 5 19
	$\cdots$	$\cdots$
	8N/(N − 4)	*24N only
$\cdots$	$\cdots$	$\cdots$
$\omega \cdot 3$	8	*24∞, *2 5 20, *2 6 12, *3 3 12, 3*12, *288, 8*2, *346, *444, 4*4, *2224, 2*24, 2 3 24, 2 4 8.

Table 18.1. (continued.)

the characteristics that arise. The finite numbers correspond to the finite groups (listed on page 58), and the first infinite number, ω, corresponds to the Euclidean plane groups, so the numbering in the table starts at $\omega + 1$.

Thurston's Geometrization Program

This work is the two-dimensional part of Thurston's geometrization program by which he planned to classify 3-manifolds by proving his Geometrization Conjecture that every 3-manifold can be cut into certain pieces having natural metrics, in an essentially unique way. This conjecture has now been proved, by Grigory Perelman [21]. The metrics are of eight different types, as compared with the three (spherical, Euclidean, hyperbolic) that happen in the two-dimensional case. But, just in that case, there is a precise sense in which almost all the manifolds are hyperbolic.

We enumerated the groups that act on two-dimensional surfaces by listing the possible orbifolds. This cannot be done in the three-dimensional case because, even after Perelman's work, we don't have a classification of even the three-dimensional closed manifolds. In a sense, Thurston's geometrization is the converse of that idea, because the information transfer is in the reverse direction: rather than deducing information about groups from information about manifolds, we deduce information about properties of possible 3-manifolds from properties of groups.

Appendix: Proof of the Inequivalence Theorem

We prove here that there are no abstract isomorphisms (other than the isotopies) among the Euclidean and hyperbolic groups with compact orbifolds—in other words, the infinite groups that do not have an ∞ in their signatures.

The proof proceeds by systematically showing that we can recover more and more geometrical information from the abstract group.

Recovering the Simplest Concepts

The only elements of finite order are reflections and rotations, so we first distinguish between these. Since the orbifold is compact, the image of a mirror line in it must also be compact, which shows that any *reflection* commutes with a translation. But the only things that can commute with a *rotation* about a point P are the elements that fix P, which generate a finite group. This starts a dictionary:

Reflections$\leftrightarrow$ Elements of order 2 with infinite centralizer.
Rotations $\leftrightarrow$ Elements of finite order and finite centralizer.

We can also tell when the mirrors of two reflections intersect— when and only when they generate a finite group. Of course the rotations that are *not* gyrations are the ones in finite groups generated by reflections, so we can identify the *gyrations* as those that aren't. The dictionary therefore continues:

Mirror line $\leftrightarrow$ Order 2 element with infinite centralizer.
Mirror lines join $\leftrightarrow$ They generate a finite subgroup.
Kaleidoscopic point$\leftrightarrow$ A maximal such subgroup.
Gyration point $\leftrightarrow$ Any other maximal finite subgroup.
Kaleidoscope $\leftrightarrow$ Maximal connected set of mirror lines.

We can also tell whether any two such things have the same type— when and only when they are conjugate—and in addition we know the degree of any gyration point or kaleidoscopic point.

Recovering the Shapes of Kaleidoscopes

The generators and relations corresponding to a kaleidoscopic part of the signature have the form

$$^{\alpha}_* {}^P \mathbf{a}^Q \mathbf{b} \ldots \mathbf{e}^S \mathbf{f}^T,$$

$$1 = P^2 = (PQ)^a = Q^2 = \ldots = S^2 = (ST)^f = T^2, \ P^\alpha = T,$$

where $P^\alpha = \alpha^{-1} P \alpha$. Then, in the doubly-infinite sequence of reflections

$$\ldots P^{\alpha^{-1}}, Q^{\alpha^{-1}}, \ldots, S^{\alpha^{-1}}, T^{\alpha^{-1}} = P, Q, \ldots, S, T = P^\alpha, Q^\alpha, \ldots,$$

we know the orders of all products of adjacent pairs, namely

$$\ldots, a, b, \ldots, e, f, a, b, \ldots, e, f, a, \ldots.$$

Let's call any reflections that satisfy these relations a *standard generating set* for the kaleidoscope they generate. Then, we'll prove that the meeting points of adjacent pairs in *any* standard generating set must actually be adjacent in the real kaleidoscope. Of course, this enables us to identify the above sequence of numbers $\ldots, a, b, \ldots, e, f, \ldots$, up to reversal, for each kaleidoscope.

The adjacency is preserved because adjacent mirrors of these reflections necessarily intersect, and so they mark out a possibly zigzag path in the real kaleidoscope (as on the left in Figure 18.4). If the meeting points are not all adjacent, this path has some extra

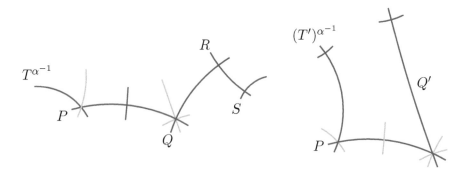

Figure 18.4. Recovering a kaleidoscope.

kaleidoscopic points along its edges; but then we can see that what it generates isn't all of the kaleidoscope.

This is particularly obvious in the general case when the kaleidoscope is a tree, because then we can find new generators $\dots, (T^{\alpha^{-1}})'$, $P, Q', R', \dots, S', T', \dots$ that "roll up" the kaleidoscope around a fundamental region for the group they generate (as on the right in Figure 18.4), where Q' is that conjugate of Q inside $\langle P, Q \rangle$ whose mirror makes the correct angle $\frac{\pi}{a}$ with that of P in the correct sense, and so on. But, it is clear from the new generators that the group they generate isn't all of the original kaleidoscope, since it still has extra kaleidoscopic points on some edges.

When there are closed polygons, we need a further argument. This is that the supposed alternative generators must also have a closed cycle, and the region enclosed by the mirrors of these must be made of several copies of the fundamental region for the real kaleidoscope, and so have area at least equal to the area of that. But, its angles must also be multiples of the angles of that region, whence (in the hyperbolic case) its area is at most that area, since it is a decreasing function of the angles. The new angles must therefore equal the old ones, so the new fundamental region is an old one, completing the proof in the hyperbolic case. In the Euclidean case, the two angle sums are necessarily equal, so the new fundamental region must be similar to the old one and if it's to generate the same group must actually be an old fundamental region.

Recovering the Topology

We have now nearly solved our problem of recovering the signature, say

$$\alpha_1 \circ^{X_1, Y_1} \dots {}^{\alpha_h} \circ^{X_h, Y_h} {}^{\beta_1} \mathbf{A}_1 \dots {}^{\beta_g} \mathbf{A}_g^{\gamma_1} *^{P_1} \mathbf{a}_1^{Q_1} \mathbf{b}_1 \dots {}^{\gamma_k} *^{P_k} \mathbf{a}_k^{Q_k} \mathbf{b}_k \dots$$

$$\dots {}^{\delta_1} \times^{Z_1} \dots {}^{\delta_x} \times^{Z_x}$$

from the abstract form of $\mathbf{G}$. We have already said enough to identify the number g of kinds of gyrations, their orders $A_1, \dots A_g$, the number k of kinds of kaleidoscopes, and their shapes $*\mathbf{a}_1 \mathbf{b}_1 \dots, \dots,$ $*\mathbf{a}_k \mathbf{b}_k \dots$, up to possible reversal of some of them.

All that's left is the topology of the orbifold, and if it is orientable, the correspondence between the two possible cyclic orders for each kaleidoscope. The topology is easy! We needn't have both handles and crosscaps, so we can suppose that one of h and x is zero. If we factor $\mathbf{G}$ by the group generated by all its gyrations and the normalizers of all its kaleidoscopes, all that's left is the group

$$\langle \alpha_1, X_1, Y_1, \ldots, \alpha_h, X_h, Y_h | \alpha_1 \alpha_2 \ldots \alpha_h = 1, \alpha_i = [X_i, Y_i] \rangle$$

in the orientable case, or

$$\langle \delta_1, Z_1, \ldots \delta_x, Z_x | \delta_1 \delta_2 \ldots \delta_x = 1, \delta_i = Z_i^2 \rangle$$

in the nonorientable one. But, the abelianized forms of these, namely

$$\mathbf{C}_\infty^{2h} \text{ and } \mathbf{C}_2 \times \mathbf{C}_\infty^{x-1},$$

are different, which determines the orientation and also determines the number h or x.

Reducing to One Last Problem

In the nonorientable case we're finished, because then the individual kaleidoscopes *can* be independently reversed. What we do in the orientable case is factor out all elements of finite order and then abelianize, obtaining the abelian group generated by

$$\overline{\alpha}_1, \overline{X}_1, \overline{Y}_1, \ldots, \overline{\alpha}_h, \overline{X}_h, \overline{Y}_h, \overline{\gamma}_1, \overline{\gamma}_2, \ldots, \overline{\gamma}_k$$

subject to the relations

$$\overline{\alpha}_1 \ldots \overline{\alpha}_h \overline{\gamma}_1 \ldots \overline{\gamma}_k = 1, \overline{\alpha}_1 = [\overline{X}_1, \overline{Y}_1], \ldots, \overline{\alpha}_h = [\overline{X}_h, \overline{Y}_h].$$

But, these imply that $\overline{\alpha}_i = 1$ and simplify the only remaining relation to $\overline{\gamma}_1 \overline{\gamma}_2 \ldots \overline{\gamma}_k = 1$. But, in this abstract group we can identify the $\overline{\gamma}_i$ up to inversion, since we have factored out the appropriate kaleidoscopic subgroup. The only thing that we don't yet know is which of $\overline{\gamma}_i$ and $\overline{\gamma}_i^{-1}$ is which. When we do know this, we'll know everything, since the shape of the corresponding kaleidoscope differs from its reversal. If we replace some of the $\overline{\gamma}_i$ by their inverses, this relation would become

$$\overline{\gamma}_1^{\pm 1} \overline{\gamma}_2^{\pm 1} \cdots \overline{\gamma}_k^{\pm 1} = 1,$$

which is equivalent to the original only when all the signs are equal.

Solving the Last Problem

Now look at (the geometrical form of) a kaleidoscope of some non-palindromic signature, such as $*2243$ (left, Figure 18.5). In the full group $\mathbf{G}$, this has four conjugacy classes of kaleidoscopic points (i.e., subgroups of a certain form), represented by

$$\mathbf{k}_0 = \langle e_0, e_1 \rangle, \quad \mathbf{k}_1 = \langle e_1, e_2 \rangle, \quad \mathbf{k}_2 = \langle e_2, e_3 \rangle, \quad \mathbf{k}_3 = \langle e_3, e_4 \rangle,$$

where e_0, e_1, e_2, e_3, e_4 correspond to consecutive edges around the appropriate boundary curve of the orbifold. Continuing along that curve would give us more such groups $\mathbf{k}_4 = \langle e_4, e_5 \rangle, \mathbf{k}_5 = \langle e_5, e_6 \rangle$, but of course these are conjugate in $\mathbf{G}$ to $\mathbf{k}_0, \mathbf{k}_1, \ldots$.

However, they give new conjugacy classes inside the proper subgroup $\mathbf{K}$ generated by all the reflections of this kaleidoscope. What happens is that the neighbors of $\mathbf{k}_2$ are alternately conjugate to $\mathbf{k}_1$ and $\mathbf{k}_3$ by e_2 and e_3, and so inside $\mathbf{K}$, and similarly the neighbors of any $\mathbf{k}_n$ are alternately conjugate inside $\mathbf{K}$ to $\mathbf{k}_{n-1}$ and $\mathbf{k}_{n+1}$, as indicated in Figure 18.5 on the right. This shows that in $\mathbf{K}$ there is a doubly infinite sequence of conjugacy classes of maximal finite subgroups generated by reflections, represented by $\ldots \mathbf{k}_{-2}, \mathbf{k}_{-1}, \mathbf{k}_0, \mathbf{k}_1, \mathbf{k}_2, \mathbf{k}_3, \ldots$. If $m \neq n$, $\mathbf{k}_m$ and $\mathbf{k}_n$ cannot be conjugate inside $\mathbf{K}$ because we can see that every reflection preserves subscripts in the figure, and therefore $\mathbf{K}$ does also.

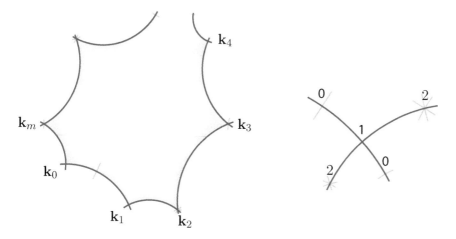

Figure 18.5. Kaleidoscope of a non-palindromic signature.

How, then, did $\mathbf{k}_4$ become conjugate to $\mathbf{k}_0$ in $\mathbf{G}$? The answer is that if the kaleidoscope corresponds to the portion

$$\gamma_{*^{e_0}} \mathbf{2}^{e_1} \mathbf{2}^{e_2} \mathbf{3}^{e_3} \mathbf{4}$$

of the signature, then γ conjugates each e_n to e_{n+4} and so each $\mathbf{k}_n = \langle e_n, e_{n+1} \rangle$ to $\mathbf{k}_{n+4} = \langle e_{n+4}, e_{n+5} \rangle$. But that refers to a particular presentation; how can we identify γ abstractly? We can't *precisely*, but we can *roughly*, because the normalizer of $\mathbf{K}$ in $\mathbf{G}$ is $\langle \mathbf{K}, \gamma \rangle$, so that the infinite cyclic group $\langle \gamma \rangle$ is determined modulo $\mathbf{K}$. Therefore, the image $\overline{\gamma}$ of γ modulo $\mathbf{K}$ is determined up to inversion, as a generator of this cyclic group.

Since the doubly infinite sequence of digits

$$\dots 2, 2, 3, 4, 2, 2, 3, 4, \dots$$

for this particular kaleidoscope is distinct from its reversal

$$\dots 3, 2, 2, 4, 3, 2, 2, 4, 3, 2, \dots,$$

we can in this case distinguish γ (or $\overline{\gamma}$) from its inverse, which decreases subscripts by 4.

Since distinguishing between the $\overline{\gamma}_i$ and their inverses was all that we had left to do, we have completed the proof that we may only reverse *all* or *none* of the kaleidoscopes.

Interlude: Two Drums That Sound the Same

When Mark Kac gave his famous lecture "Can You Hear the Shape of a Drum?" [20] in 1965, he popularized an old and important problem, whose technical statement is, "Does the Laplace spectrum of a plane region determine its shape?" The Laplace spectrum consists of the eigenvalues of the Laplacian operator ∇^2. The answer (which is "no!") was finally found by Gordon, Webb, and Wolpert [14] (relying heavily on a crucial contribution by Sunada [27]) only in 1992, about a century after the problem was first posed.

The Gordon-Webb-Wolpert examples were rather complicated. The simplest known examples and proof were found by Buser, Conway, Doyle, and Semmler [3]. The book *The Sensual Quadratic Form* [9] gives a simple exposition of their proof of the following result:

Take an acute angled scalene triangle. Create a copy of it, and reflect both copies once across each edge. On one copy, further reflect across the right edges of the images; on the other copy reflect instead across the left edges. Each resulting set of seven triangles is a drum, the two drums have different shapes, and they sound the same.

The shapes of the triangles here are totally immaterial to the argument—the two drums will be isospectral no matter what shape they are, and they will usually not be isometric. Indeed, the triangles can be replaced by arbitrary curved surfaces (maybe with holes!) provided only that adjacent ones are related by reflection. To explain how the two propeller shapes were found, we study the effect of making the basic triangle be the equilateral hyperbolic triangle that is the orbifold of ∗**444** (even though this is one of the infinitely rare cases that yields isometric drums).

Then (Figure 18.6, right) the two drums are orbifolds for two mirror image copies of the hyperbolic group ∗**424242**. Since the reflections that generate these groups all belong to the ∗**444** group defined by the central triangle, this shows that ∗**444** has two distinct subgroups of index 7 and signature ∗**424242**. It turns out that the intersection **I** of all the conjugates

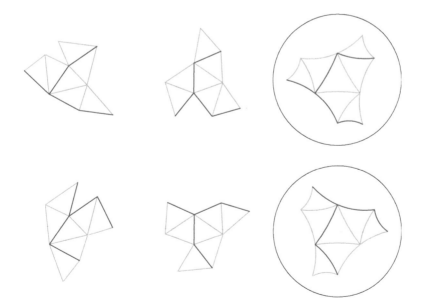

Figure 18.6. The first column shows a pair of drums obtained from one shape of triangle; the middle column shows an alternative pair obtained from another shape, and the right column shows hyperbolic drums.

of these groups has index 24 in each of them, and so index 168 in the central
∗**444** group. Its signature is $\times^{23}$. We can at least check this using the index
rules:

Group	Cost	Characteristic	Index
∗**444**	$1 + \frac{3}{8} + \frac{3}{8} + \frac{3}{8} = 2\frac{1}{8}$	$-\frac{1}{8}$	
			7
∗**424242**	$1 + \frac{3}{8} + \frac{1}{4} + \frac{3}{8} + \frac{1}{4} + \frac{3}{8} + \frac{1}{4} = 2\frac{7}{8}$	$-\frac{7}{8}$	
			24
$\times^{23}$	$23 = 2 + 21$	-21	

Indeed, it is not too hard to convert this check into a proof. For,
obviously, there is a conjugate of either group that avoids any particular
reflection or rotation, which entails that we can take the signature of **I** to
consist entirely of either $\times$'s or $\bigcirc$'s; but we cannot get the correct characteristic -21 using only $\bigcirc$'s. We defined **I** in such a way that it is obviously
a normal subgroup of all the other groups mentioned. The group ∗**444**
maps modulo **I** to a group of order 168 and the two groups ∗**424242** to
two order-24 subgroups of this.

This explains how the example was found. The group of order 168 is the
well-known group that consists of all automorphisms of the finite projective
plane with seven points and seven lines, and the two subgroups of index 7
are the stabilizers of a point and line, respectively. The crucial fact about
these groups is that they are themselves *isospectral* in the sense that they
contain the same number of elements in each conjugacy class of the group
of order 168. Sunada's theorem tells us that isospectral subgroups of the
group with a given orbifold will have isospectral orbifolds.

Many other examples can be found by the same method. What one has
to do is find a finite group, with two subgroups isospectral in this group
theoretical sense, that acts on some surface in such a way that the orbifolds
of those subgroups are topological disks.

Homophonic Drums

Peter Doyle has observed that in both the original example of Gordon-
Webb-Wolpert and these simple propeller examples, the two drums don't
really "sound the same." The argument proves only that they have the
same resonant frequencies, which just entails that if you place them in
a hall with an orchestra, they will respond in the same way. They only
"*resound* the same"!

The two "peacocks" of Doyle and Conway cope with this problem.
(Figure 18.7). Each is made of 21 copies of the same triangle and has a
unique internal "node" at which six of those triangles meet. The proof
in [3] shows that the sound produced by hitting the node of a drum in

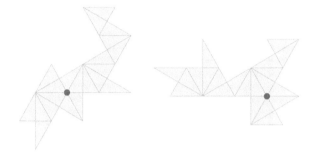

Figure 18.7. Peacocks rampant and couchant.

the shape of the "peacock rampant" is exactly the same as that produced by hitting the node of a drum in the shape of the "peacock couchant." These were obtained by Sunada's method from a suitable mapping from ∗**633** to the automorphism group of the 21-point projective plane. The signatures of the appropriate pair of index 21 subgroups are ∗**63623333** and ∗**66323333** and that of the intersection of all their conjugates is $\times^{1682}$, which you can check must have index 20,160 in ∗**633** and index 960 in the above two subgroups.

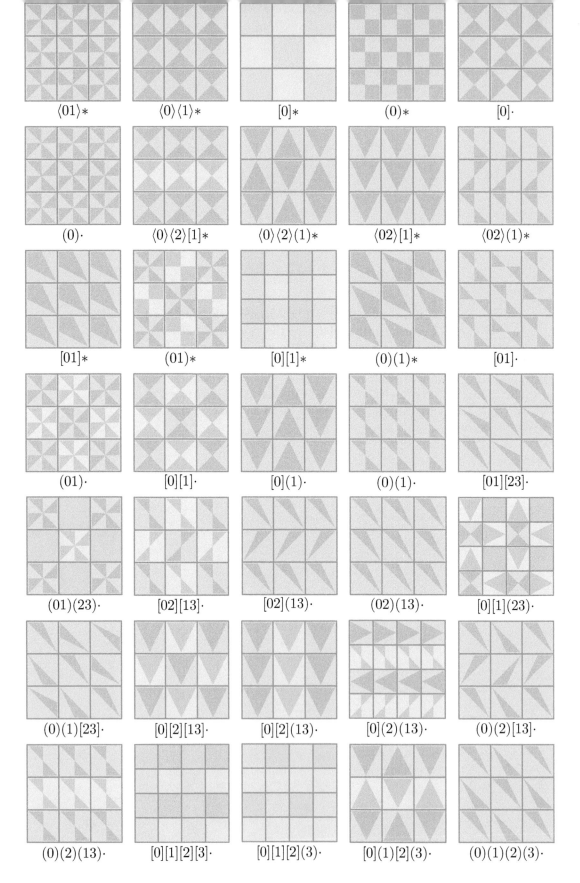

$\langle 01\rangle *$ $\langle 0\rangle\langle 1\rangle *$ $[0]*$ $(0)*$ $[0]\cdot$

$(0)\cdot$ $\langle 0\rangle\langle 2\rangle[1]*$ $\langle 0\rangle\langle 2\rangle(1)*$ $\langle 02\rangle[1]*$ $\langle 02\rangle(1)*$

$[01]*$ $(01)*$ $[0][1]*$ $(0)(1)*$ $[01]\cdot$

$(01)\cdot$ $[0][1]\cdot$ $[0](1)\cdot$ $(0)(1)\cdot$ $[01][23]\cdot$

$(01)(23)\cdot$ $[02][13]\cdot$ $[02](13)\cdot$ $(02)(13)\cdot$ $[0][1](23)\cdot$

$(0)(1)[23]\cdot$ $[0][2][13]\cdot$ $[0][2](13)\cdot$ $[0](2)(13)\cdot$ $(0)(2)[13]\cdot$

$(0)(2)(13)\cdot$ $[0][1][2][3]\cdot$ $[0][1][2](3)\cdot$ $[0](1)[2](3)\cdot$ $(0)(1)(2)(3)\cdot$

– 19 –

Archimedean Tilings

Archimedes enumerated the convex polyhedra with regular faces and only one type of vertex (because the symmetry group is transitive on the vertices). Many people have independently enumerated the tilings of the Euclidean plane that satisfy the same condition, and a few have tried to enumerate the corresponding tilings of the hyperbolic plane and achieved some partial results.

Often there is some subgroup **H** of the full symmetry group **G** of such a tiling that remains transitive on the vertices; in this case we say that the tiling is *Archimedean relative to* **H**, to contrast it with the *absolute* case when **H** = **G**.

For example, the Archimedean *snub square tessellation* (top right) has full symmetry group **4∗2**, with respect to which it is absolute. However, coloring the tiles as in the second marginal figure, we see that the group **442** still acts transitively on its vertices, and so it is Archimedean relative to this subgroup.

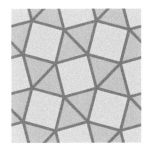

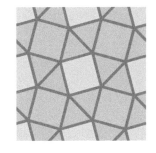

The complete classification of all Archimedean tilings, both relative and absolute, appears for the first time in this book. As usual the guiding idea is that of orbifold, since once we have specified the orbifold we can recover the tiling by "unwrapping." The orbifold of an Archimedean tiling is hard to visualize, in view of all those faces. Fortunately, we don't have to visualize the faces, because we can remove them from the tiling with no loss of information if we're careful.

What we do is enlarge the vertices to blobs and the edges joining them to strips by slightly shrinking the faces. (One can think of these blobs and strips as the grout between the tiles.) Then we simply remove these shrunken faces! The result is like a doily made of paper

---

(opposite page) The thirty-five relative Archimedean tilings of the Euclidean plane by squares.

circles and strips. This doily still contains all the information, since we can recover the original tessellation simply by sewing polygonal patches onto all the boundary curves.

The orbifold of the doily is really quite a simple thing, since the Archimedean property of the tessellation implies that it consists of just one blob, possibly folded and probably connected to itself by various strips, again possibly folded.

Below we see the doily for the snub square tessellation, with its full symmetry group **4∗2**; the blob in it has five *arms* which are numbered 0, 1, 2, 2′, and 1′, because in the orbifold arms 1′ and 2′ have been folded onto 1 and 2, respectively. Also, arm 0 has been folded onto itself to become a *half arm* ⟨0⟩. Finally, the outer ends of arms 1 and 2 have been identified, turning those arms into a *band* (12) that joins the blob to itself.

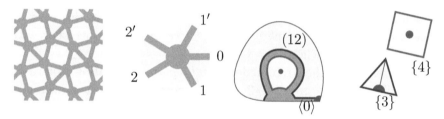

What has become of the two types of faces? The square face has become the brown boundary curve, which now consists of just one edge in the orbifold; since this goes around the order-4 gyration point, we know it must have had four sides in the original, as indicated by our {4}. The triangle has become the purple boundary, which has $1\frac{1}{2}$ sides (the full side at arms 1 and 2 and the half side at 0). But, since it ends on the red mirror line, its original polygon was a {3}.

We can convey the topological information here by the *permutation symbol* ⟨0⟩(12), (which conveys the topology), together with the *face-code* 34343, starting with the face between arms 0 and 1.

We combine both parts into a single symbol, here

$$\langle 0 \rangle (12) * (3, 4, 3, 4, 3),$$

with the separating symbol (here ∗) indicating the local symmetry at the vertex.

The Permutation Symbol

In general, neighboring blobs in an archimedean tiling can be connected to one another by arms in just a few ways.

Suppose that arm i of one blob is connected to arm i of another vertex. Then there is a symmetry fixing the central point of this edge. If there is only a rotational symmetry, of order 2, then in the orbifold we see the arm grasp an order 2 cone point: we call this a *rotary arm* (i). If there is a reflection symmetry interchanging the two blobs, in the orbifold we see the arm run from the vertex to a boundary of the orbifold; this is a *folded band* $[i]$. Finally, if there is a symmetry $*\mathbf{2}$ at the center of this edge, in the orbifold we have a *half arm* $\langle i \rangle$, just as we saw in the snub square tessellation. The type of brackets used here hints at the shape at the end of the arm.

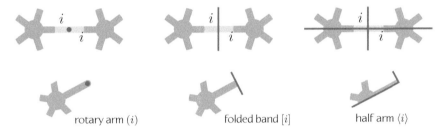

rotary arm (i) folded band $[i]$ half arm $\langle i \rangle$

Or, we might have arm i of a blob connected to arm $j \neq i$ of another blob. If there is a symmetry interchanging the two sides of this edge in the tiling, we have a *half band* $\langle ij \rangle$, running along a boundary of the orbifold. If not, and the blobs are of the same orientation, we see an *untwisted band* (ij) in the orbifold, just as in the snub square tessellation. Finally, if the blobs are of opposite orientations, the orbifold will contain a *twisted band* $[ij]$.

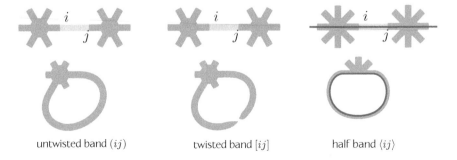

untwisted band (ij) twisted band $[ij]$ half band $\langle ij \rangle$

The parts of the orbifold that correspond to the faces we removed must each be disks; these may contain at most one cone point or kaleidoscopic point according to whether they lie in the interior of the orbifold or on its boundary, as shown on the left in the figure below.

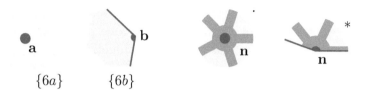

$$\{6a\} \qquad \{6b\}$$

Our separating symbol is a · if the blob lies in the interior of the orbifold, when it might become a cone point of order **n**, and a ∗ if it lies on a boundary of the orbifold, when it might become a kaleidoscopic point ∗**n**.

The key observation is that an arbitrary permutation symbol *precisely* describes the topology of the orbifold, up to the orders of cone and kaleidoscopic points! This is simply because there is just one way to zip disks to the boundary of the doily to get a closed surface.

The orbifold for the symbol $[0](12)[34]\cdot$ is drawn below: we see that the blob may have a cone point of order $\mathbf{n} \geq 1$ at its center. There are two boundary curves of the doily, one of which must meet the boundary of the orbifold. Consequently, the orbifold has exactly two faces, one with (perhaps) a cone point of order **a** and the other with (perhaps) a kaleidoscopic point of order **b**. We shall see that the values of **a**, **b**, and **n** can be deduced from the face code.

Euler Characteristic

If a filled in doily has f faces and g bands (of all kinds—untwisted, twisted, half, and folded), the Euler characteristic is $1 + f - g$. Half arms and rotary arms contribute nothing. This number, together with the number of boundaries and the orientability of the orbifold, determines the orbifold's topology.

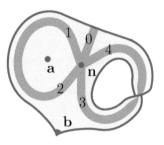

$$[0](12)[34]\cdot(8b, a, 8b, 8b, 8b)^n$$

This orbifold is nonorientable with one boundary component; its Euler characteristic is 0, so the orbifold must be a Möbius band and the tiling must have symmetry $\mathbf{a}*\mathbf{b}\times$. Moreover, counting the edges of the two faces, we see that the face code can only be $(8a, 8a, 8a, b, 8a)^n$. Conversely, given the face code, we can deduce $\mathbf{a}$, $\mathbf{b}$, and $\mathbf{n}$, and so know the symmetry.

Consider the more complicated example

$$\langle 08 \rangle (1)[26](34)(57)*.$$

What is the corresponding orbifold? Beginning at arm 0 and tracing around, we find that the doily has two boundaries, colored purple and yellow, and so the orbifold will be filled in with two faces. It's difficult to draw these, though!

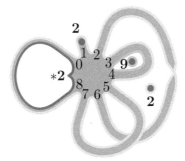

$$\langle 08 \rangle (1)[26](34)(57)*(14^3, 9, 14^8, 9, 14^3)^2$$

What can we deduce about the local features of the orbifold from the symbol? There is an order-2 cone point at the end of rotary arm (1). The boundary of the half band might have a kaleidoscopic point $\mathbf{n}$ at the blob, and two faces may each have cone points of orders, say, a for yellow and b for purple. The yellow face has seven edges and so will be a $(7a)$-gon in the tiling. The purple face has just one edge and so will be a b-gon in the tiling. Any tiling with the permutation symbol $\langle 08 \rangle (1)[26](34)(57)*$ must therefore have a face code of the form $((7a)^3, b, (7a)^8, b, (7a)^3)^n$; this immediately implies that for the face code we were given, $(14^3, 9, 14^8, 9, 14^3)^2$, we have $n = 2$, $a = 2$, and $b = 9$.

What is the orbifold's topology? The orbifold is nonorientable because it includes the twisted band [26]. The orbifold has one boundary along the half band $\langle 08 \rangle$. There are four bands altogether, so the Euler characteristic is $1 + f - g = 1 + 2 - 4 = -1$, and the underlying surface of the orbifold is $*\times\times$. We now know the symmetry of the tiling with this symbol: it has signature $\mathbf{229}*\mathbf{2}\times\times$.

Existence

Is there really a tiling with this face code? We are actually asking whether there exist a regular 9-gon and a regular 14-gon in the hyperbolic plane, with matching edge-lengths so that four of the 9-gons and twenty-eight of the 14-gons can fit together at a vertex.

The answer is easy: yes, there is, a unique one with the given symbol. Why is this? The answer is that the universal cover of the orbifold is a plane that is topologically tiled in the required way, and that there is a unique edge-length for which the angles of the corresponding regular polygons that should fit around a vertex will add to $360°$. To see this, it suffices to note that when the edge-length is very small the angles will be close to their values in the Euclidean case, which add to more than $360°$, and that they decrease to zero as the edge length increases to ∞.

With trivial modifications the argument works also in the Euclidean and spherical cases, with the single exception that if the group signature derived from the symbol is one of the bad ones ($\mathbf{ab}$ or $*\mathbf{ab}$ with $a \neq b$), then there is no orbifold and so no tessellation.

Relative versus Absolute

Our combined symbol exactly classifies tilings that are Archimedean relative to $\mathbf{H}$, that is, pairs $(T, \mathbf{H})$ where T is a tiling by regular polygons and $\mathbf{H}$ is a group of automorphisms of T that is transitive on the vertices.

A tiling with permutation symbol $[0][12]*$, for example, cannot be absolute. Its orbifold has only one face, so the tiling is regular and there is more symmetry than the symbol reflects. As a regular tiling, it has a different symbol $\langle 0 \rangle *$ and orbifold.

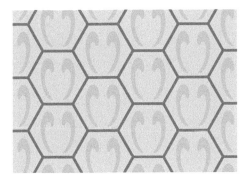

Figure 19.1. The tiling as drawn has less symmetry than the tiling by unmarked, uncolored polygons. Accordingly, the permutation symbol corresponding to this tiling, [0][12]∗, can only be relative, not absolute.

Again, the doily for $(0)(1)(2)\cdot$ can only be relative, since it has only one face and so describes a regular tiling; the absolute symbol will be $\langle 0 \rangle *$. The more complicated example of $(03)(16)(25)(47)\cdot$ has two faces but has a rotational symmetry and is also necessarily relative.

The symbol $[0][1][2]\cdot$ describes both relative and absolute tilings. The orbifold has three faces, each with a kaleidoscopic point, of orders a, b, and c. If all three of these orders are distinct, then there is no symmetry of the doily and the tiling is absolute; otherwise, there is such a symmetry and the tiling is merely relative.

A tiling T will automatically be absolute with respect to the full group **G** of *all* its automorphisms. So, we can classify the *absolute* Archimedean tilings by discarding any for which $\mathbf{G} \neq \mathbf{H}$.

For the small cases it is easy to apply tests like those we just described to decide whether a symbol is merely relative, but what should we make of

$$[0][15][2][37](4)(6)\cdot \text{ or } [0](1)[25][3](47)(6)\cdot \text{ ?}$$

Both of these have more than one face and neither has an obvious symmetry of the doily. A simple algorithm quickly shows that both are relative.[1]

---

[1]The simple tests do settle the matter for the vast majority of symbols: these two examples are the smallest for which more subtle arguments are needed.

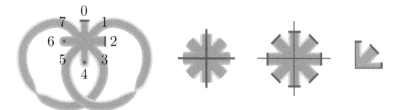

[0][15][2][37](4)(6)·, shown at left, is relative. The group *2 acts on the vertex. The reflections preserving 4 and 6 transmute (4) and (6) into [4] and [6]. The rotation swapping 1 and 5, and 3 and 7, transmutes [15] into [1][5] and [37] into [3][7], yielding the doily [0][1][2][3][4][5][6][7]· with face code (*abbaabba*), which is symmetric under *2. The absolute code is thus ⟨0⟩[1]⟨2⟩*.

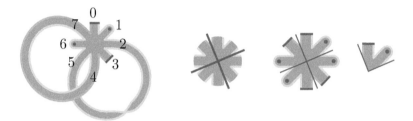

[0](1)[25][3](47)(6)·, shown at left, is also relative. The group *2 acts on the vertex; the reflection swapping 2 and 5 transmutes [25] into (2)(5); the reflection swapping 4 and 7 transmutes (47) into [4][7], yielding the doily [0](1)(2)[3][4](5)(6)[7]· with face code (*aaabaaab*). This is symmetric under *2, and the absolute code is [0](1)*.

To determine that a symbol is merely relative we first must see that the arrangement of faces at a vertex has greater symmetry than implied by the symbol. This was just the test that showed that [0][1][2] could not be relative for distinct a, b, and c. It is easily made algorithmic, but we shall merely suggest how by the discussions of the two examples in the figures above.

Replace:	With:	If There Is:
[ij]	(ij)	just a reflection preserving i and j
[ij]	(i)(j)	just a reflection swapping i and j
[ij]	[i][j]	just a rotation swapping i and j
[ij]	[i][j]	all of the above
(ij)	(ij)	just a reflection preserving i and j
(ij)	[i][j]	just a reflection swapping i and j
(ij)	(i)(j)	just a rotation swapping i and j
(ij)	[i][j]	all of the above
(i)	[i]	a reflection preserving i
[i]	[i]	a reflection preserving i

Enumerating the Tessellations

We shall now use this theory to enumerate our Archimedean tessellations of small rank. The *rank* is the number of half or halved arms at a vertex.

In the following tables we first list all very small symbols with rank up to 3. We next list all absolute symbols with rank up to 5. We then list all absolute examples for which the vertex is in the interior of the doily, up to rank 10, which in this case (since there are no half arms) means valence 5. Finally, at the end of this chapter, we give a table listing the number of absolute and relative symbols up to rank 19.

We write the rank as a sum of 2's (for full arms) and 1's (for half arms) and follow it by $(*)$ or $(\cdot)$ according as there is or is not a reflection through the vertex. We see that only three of the first nine cases can be absolute.

Rank	Group	Symbol	Comment	Example
$1(*)$	$*\mathbf{na2}$	$\langle 0 \rangle *(a)^n$	always absolute	all regular tilings
$2(\cdot)$	$\mathbf{na2}$	$(0)\cdot(a)^n$	always relative	
$2(\cdot)$	$\mathbf{n*a}$	$[0]\cdot(2a)^n$	always relative	
$2(*)$	$\mathbf{2*na}$	$(0)*(2a)^n$	always relative	
$2(*)$	$*\mathbf{nab}$	$[0]*(a,b)^n$	absolute if $a \neq b$	semiregular tilings
$(1{+}1)(*)$	$*\mathbf{n2a2}$	$\langle 0 \rangle \langle 1 \rangle *(2a)^n$	always relative	
$(1{+}1)(*)$	$\mathbf{a*n}$	$\langle 01 \rangle *a^n$	always relative	
$(2{+}1)(*)$	$\mathbf{2*na2}$	$\langle 0 \rangle (1)*(3a)^n$	always relative	
$(2{+}1)(*)$	$*\mathbf{n2ab}$	$\langle 0 \rangle [1]*(2a,b,2a)^n$	absolute if $2a \neq b$	truncated regular tilings

Now we give a table up to rank 5 for the absolute cases only.

Rank	Group	Symbol	Condition
$1(*)$	$*\mathbf{na2}$	$\langle 0 \rangle *(a)^n$	—
$2(*)$	$*\mathbf{nab}$	$[0]*(a,b)^n$	$a \neq b$
$(2{+}1)(*)$	$*\mathbf{n2ab}$	$\langle 0 \rangle [1]*(2a,b,2a)^n$	$2a \neq b$
$(2{+}2)(*)$	$\mathbf{2*nab}$	$(0)[1]*(3a,3a,b,3a)^n$	$3a \neq b$
$(2{+}2)(*)$	$*\mathbf{nabc}$	$[0][1]*(a,2b,c,2b)^n$	$a \neq c$
$(1{+}2{+}1)(*)$	$*\mathbf{n2ab2}$	$\langle 0 \rangle [1] \langle 2 \rangle *(2a,2b,2b,2a)^n$	$a \neq b$
$(1{+}2{+}2)(*)$	$\mathbf{2*n2ab}$	$\langle 0 \rangle (1)[2]*(4a,4a,b,4a,4a)^n$	$4a \neq b$
$(1{+}2{+}2)(*)$	$\mathbf{2*n2ab}$	$\langle 0 \rangle [1](2)*(2a,3b,3b,3b,2a)^n$	$2a \neq 3b$
$(1{+}2{+}2)(*)$	$*\mathbf{n2abc}$	$\langle 0 \rangle [1][2]*(2a,2b,c,2b,2a)^n$	$2a,2b,c$ not all equal
$(1{+}2{+}2)(*)$	$\mathbf{a*n2}$	$\langle 0 \rangle (12)*(3b,2a,3b,3b,2a)^n$	$2a \neq 3b$

We now consider the absolute cases for which there is no collapse of the vertex, up to rank 10; the rank is then twice the valence.

Valence	Symbol	Conditions
3	$[0][1][2]\cdot(2a,2b,2c)$	a,b,c distinct
4	$[0](1)(23)\cdot(6a,6a,b,6a)$	$6a \neq b$
	$[0][1][2][3]\cdot(2a,2b,2c,2d)$	see † below
	$[0][1][2](3)\cdot(2a,2b,4c,4c)$	$a \neq b$
5	$[0][12](34)\cdot(8a,8a,8a,b,8a)$	$8a \neq b$
	$[0](12)(34)\cdot(6a,b,6a,c,6a)$	$b \neq c$
	$(0)[12](34)\cdot(4a,4a,4a,b,4a)$	$4a \neq b$
	$(0)(12)(34)\cdot(3a,b,3a,c,3a)$	$b \neq c$
	$[0][1][2][34]\cdot(2a,2b,6c,6c,6c)$	$a \neq b$
	$[0][1][2](34)\cdot(2a,2b,4c,d,4c)$	$a \neq b$
	$[0][1](2)[34]\cdot(2a,8b,8b,8b,8b)$	$a \neq 4b$
	$[0][1](2)(34)\cdot(2a,6b,6b,c,6b)$	$2a,6b,c$ not all equal
	$[0][1][3][24]\cdot(2a,4b,4c,4b,4c)$	$b \neq c$
	$[0](1)(2)(34)\cdot(8a,8a,8a,b,8a)$	$8a \neq b$
	$[0](1)[3][24]\cdot(6a,6a,4b,6a,4b)$	$3a \neq 2b$
	$[0](1)[3](24)\cdot(6a,6a,4b,4b,6a)$	$3a \neq 2b$
	$[0](1)(3)(24)\cdot(6a,6a,2b,2b,6a)$	$3a \neq b$
	$[0][1][2][3][4]\cdot(2a,2b,2c,2d,2e)$	see ‡ below
	$[0][1][2][3](4)\cdot(2a,2b,2c,4d,4d)$	$a \neq c$
	$[0][1][2](3)(4)\cdot(2a,2b,6c,6c,6c)$	$a \neq b$
	$[0][1](2)[3](4)\cdot(2a,4b,4b,4c,4c)$	$b \neq c$

† No symmetry of a square with vertices labeled a, b, c, d; that is, $a \neq c$, and $b \neq d$ and neither $a = b$ and $c = d$, nor $a = d$ and $b = c$.

‡ No symmetry of a pentagon with vertices labeled a, b, c, d, e.

Archimedes Was Right!

We have done enough to verify Archimedes' assertion that the only finite "Archimedean" polyhedra are precisely those illustrated in the first half of Table 19.1. Similarly, the Archimedean plane tessellations are those illustrated in the second half of Table 19.1.

Since the angle of a regular polygon is at least 60°, there can be at most five around a vertex, making the rank at most 10. If there is no collapse at the vertex the polyhedron is included in the latter part of our table, and if there is it is in the former part, since then its rank is at most 5. All we have to do is pick out the cases for which the group is spherical, which is easy.

The enumeration of Euclidean Archimedean tilings is almost the same. The valence is at most 6 and can only be 6 for the regular triangular tessellation itself, so it is still true that all cases are in our table; the answers are found by picking out those whose group is Euclidean.

The Hyperbolic Archimedean Tessellations

The remaining tessellations are all hyperbolic. The number of permutation symbols increases tremendously with rank. Also, for any given permutation symbol, there are infinitely many distinct possible face codes.

We list the number of distinct symbols with a given rank. A permutation symbol is listed as absolute if it describes some absolute tilings. In the headings $a(n)$ indicates the total number of absolute symbols of rank n and $r(n)$ the total number of all symbols, relative and absolute, of rank n. We further divide these into $\cdot a(n)$ and $\cdot r(n)$, for which the vertex of the doily lies in the interior of the orbifold— and so the valence is $n/2$— and $*a(n)$ and $*r(n)$, for which the vertex lies on a mirror.

$n/2$	$a(n)$	$r(n)$	$\cdot a(n)$	$\cdot r(n)$	$*a(n)$	$*r(n)$
1/2	1	1			1	1
1	1	6	0	2	1	4
3/2	1	2			1	2
2	3	14	0	5	3	9
5/2	4	6			4	6
3	12	32	1	8	11	24
7/2	16	20			16	20
4	48	100	3	24	45	76
9/2	64	76			64	76
5	210	324	17	52	193	272
11/2	276	312			276	312
6	946	1285	74	185	872	1100
13/2	1252	1384			1252	1384
7	4510	5442	343	578	4167	4864
15/2	6023	6512			6023	6512
8	22380	25692	1593	2412	20787	23280
17/2	30364	32400			30364	32400
9	116481	128354	7797	10082	108684	118272

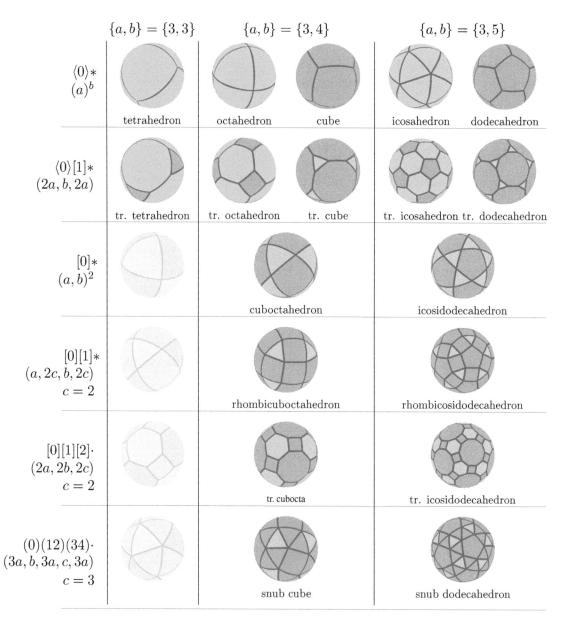

Table 19.1. The Archimedean polyhedra and tessellations. The spherical and Euclidean Archimedean tilings are shown. Each absolute tiling is shown with its name, "tr." being short for "truncated." The relative tilings are lightened. With the single exception of "deltille," the valence of an Archimedean spherical or Euclidean tiling can be at most five, and so we can be sure their enumeration is complete by examining our enumeration of small symbols.

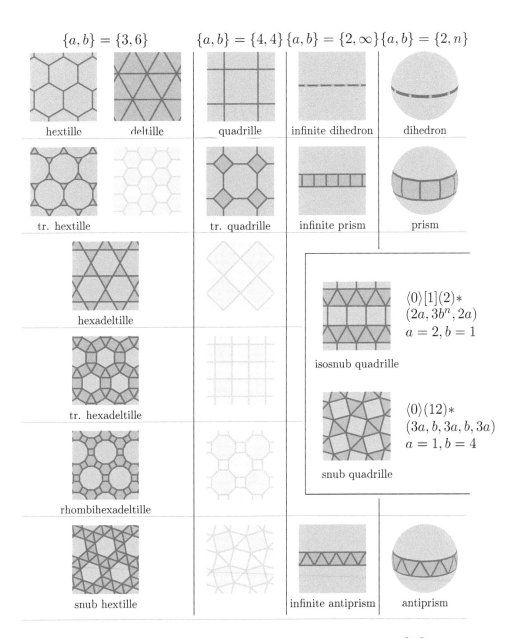

Table 19.1. (continued.) Allowing infinite faces—kaleidoscopic or gyration points of infinite order—we gain the infinite prism, the infinite antiprism, and the infinite dihedron. The gaps on this side of the table would be filled by tilings that include two-sided polygons of no area, which we do not allow.

Examples and Exercises

Try your hand at verifying these symbols.

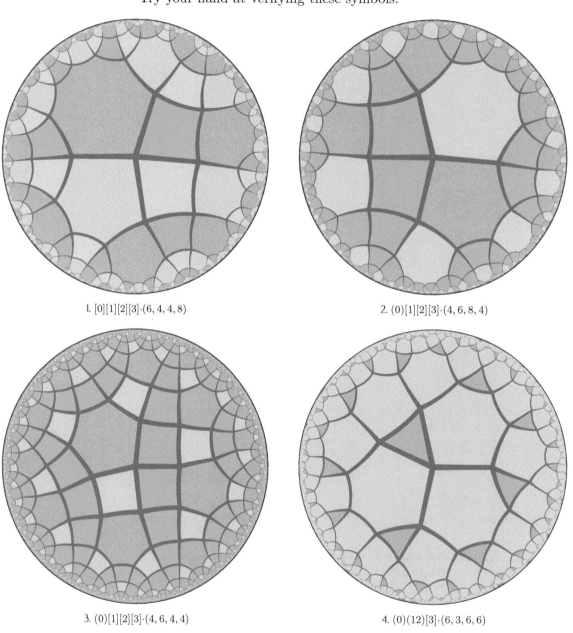

1. $[0][1][2][3] \cdot (6,4,4,8)$

2. $(0)[1][2][3] \cdot (4,6,8,4)$

3. $(0)[1][2][3] \cdot (4,6,4,4)$

4. $(0)(12)[3] \cdot (6,3,6,6)$

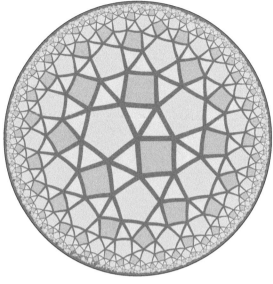

5. $(0)(12)(34)\cdot(3,4,3,5,3)$

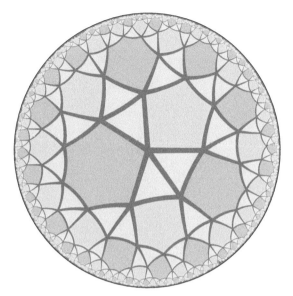

6. $(0)(12)(34)\cdot(3,5,3,6,3)$

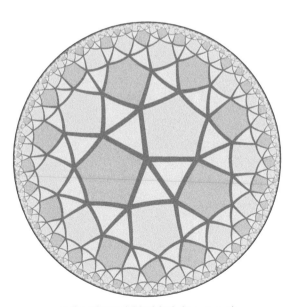

7. the relative $(0)(12)(34)\cdot(3,5,3,5,3)$

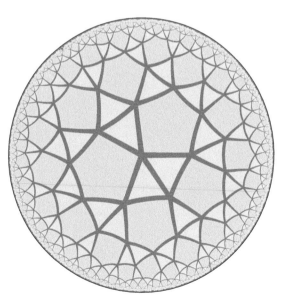

8. the absolute $\langle0\rangle(12)*(3,5,3,5,3)$

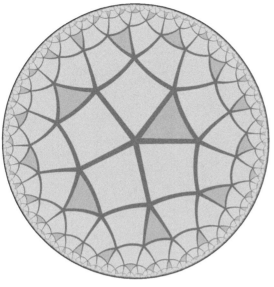

9. $(0)(12)[34]\cdot(3,4,4,4)$

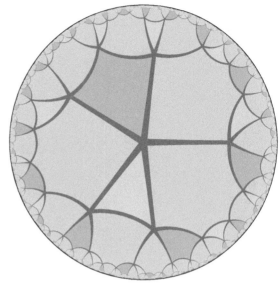

10. $(0)[1][2](34)\cdot(6,4,6,3,6)$

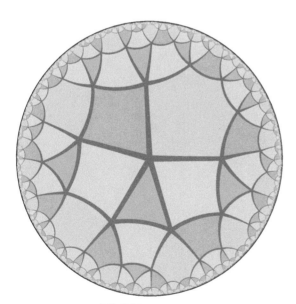

11. $[0][1][2](34)\cdot(6,4,4,3,4)$

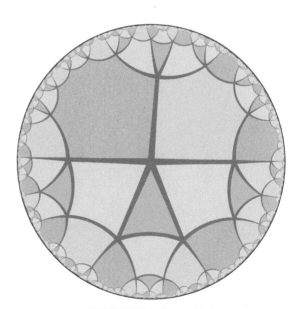
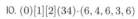

12. $[0][1][2](34)\cdot(6,8,4,3,4)$

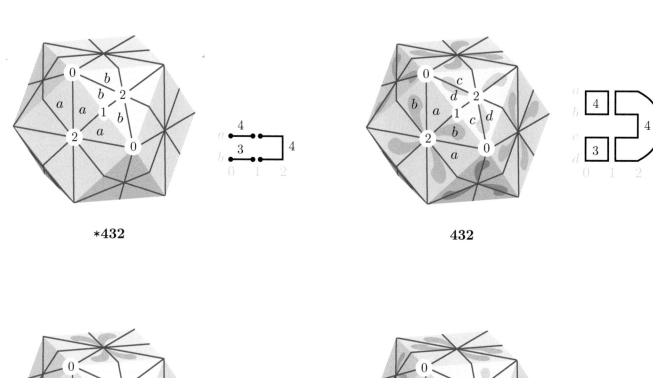

*432

432

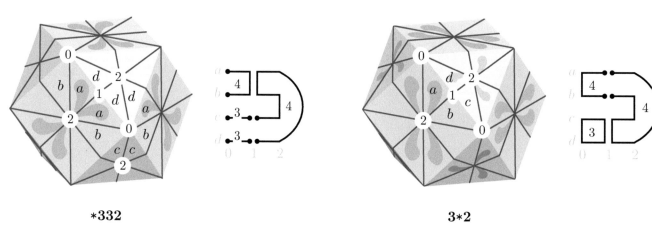

*332

3*2

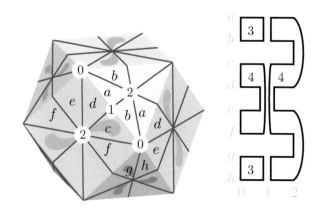

332

– 20 –

Generalized Schläfli Symbols

The ordinary Schläfli symbol, $\{p, q\}$ in Coxeter's version (Schläfli wrote (p, q) and Hoppe $(p|q)$), denotes the Platonic solid whose faces are regular p-gons and whose vertex figures are regular q-gons. In this chapter, we describe a generalization we call the "generalized Schläfli symbol of Dress and Delaney" that works for all topologically spherical polyhedra (and as we shall see later, also in higher dimensions). Andreas Dress, who really coined the symbol, called it the "Delaney Symbol" because he got the idea from a paper of M. Delaney. Our changed name reflects the fact that the version we present here visibly generalizes the Schläfli symbol.

Flags and Flagstones

Mathematicians use the word "flag" for a collection of mutually incident spaces, one of each dimension up to some limit—our little picture shows why! The flags of a polyhedron P are those whose points, lines, and planes are chosen from the vertices, edges, and faces of the polyhedron P.

For our purposes, it is useful to note that these flags correspond to the tiles of the *barycentric subdivision* of P, which we therefore call "flagstones."

This subdivision is obtained by dividing each n-gonal face of P into $2n$ triangles whose new vertices are at the center of that face and its edges. We shall number each vertex with the appropriate dimension: 0 if it is a vertex of the original solid, 1 if it's the midpoint of an edge, and 2 if it's the center of a face.

A flag is a collection of a point, a line, a plane, ...

(opposite page) The Schläfli symbols for the cuboctahedron under its full group of symmetries, the three subgroups of index 2 in that, and the one of index 4.

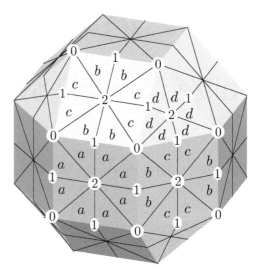

The figure above shows the barycentric subdivision of the rhombicuboctahedron, which has three types of face: equilateral triangles, regular squares, and half-regular squares. Under the symmetries of this polyhedron, its flagstones (or flags) fall under four types:

a: those in the regular square faces,

b: those next to type a, but in half-regular squares,

c: those in half-regular squares next to the triangular faces,

d: those in the triangular faces.

For a given dimension number d, we say one flagstone is d-joined to another if they share all their vertices except those numbered d. In the rhombicuboctahedron, a flagstone of type a is 2-joined to one of type b, but 0-joined and 1-joined to ones of type a.

The (generalized) Schläfli symbol has a row for each type of flagstone and a column for each of the three dimensions, 0, 1, and 2. Each row has horizontal lines joining each dimension to the next, while the dth column contains vertical lines that indicate the d-joins.

What we have just described would, for the rhombicuboctahedron, produce the figure in Figure 20.1(b), but it makes things very much clearer if we double everything for the intermediate dimensions

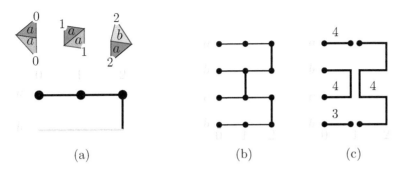

Figure 20.1. (a) The row for the flagstones of type a: they are 0-connected and 1 connected to a, and 2-connected to b. (b) The full Schläfli symbol for the rhombicuboctahedron. (c) The clearer symbol produced by doubling everything for dimension 1: we add numbers to describe the size of each face and vertex figure.

(here only in dimension 1) as in Figure 20.1(c), and we shall always do this in the future.

Each component of Figure 20.1 so modified now receives a number. For an ordinary polyhedron, these numbers are, between dimensions 0 and 1, the sizes of the corresponding faces and, between dimensions 0 and 2, the sizes of the vertex figures.

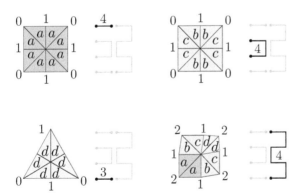

How the numbering arises.

Before we give more precise and more general definitions, let's give some properties of this new kind of Schläfli symbol. The most important one is that by deleting the parts corresponding to a given dimension, one obtains the (generalized) Schläfli symbols of the faces of that dimension. So, for instance, covering up the right-hand side

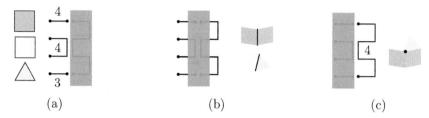

Figure 20.2. The effects of covering up various parts of the figure: (a) we see there are three kinds of faces, two of which are regular and one of which is half regular; (b) we see there are two kinds of edges; (c) we see there is just one kind of vertex, which is just $\frac{1}{4}$-regular.

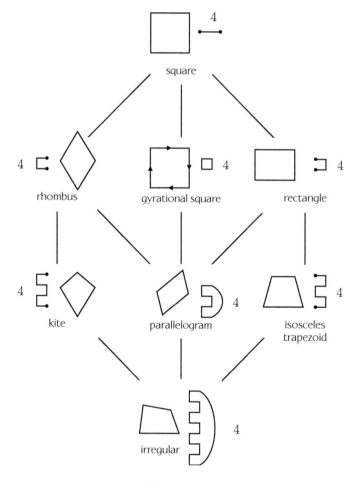

Figure 20.3. Schläfli symbols of quadrilaterals.

of our example (as in Figure 20.2(a)), we see that the rhombicuboctahedron does indeed have three types of face: two four-sided ones and one three-sided one.

Moreover, it actually tells us that the three-sided face is regular as is one of the two kinds of four-sided faces, while the other kind of four-sided face is only half regular. This is because the generalized Schläfli symbol of the regular polytope $\{p, q, r, ...\}$ just has one row:

$$\overset{p}{\bullet\!\!-\!\!\bullet}\overset{q}{\bullet\!\!-\!\!\bullet}\overset{r}{\bullet\!\!-\!\!\bullet}\ \cdots$$

Correspondingly, a polytope that is only half regular will have two rows. In general, the number of rows in the Schläfli symbol is, of course, the number of orbits of its flags (or flagstones), which is called the *flag rank*. If the flag rank is r, we sometimes say that the polytope is $\frac{1}{r}$-regular.

Covering up the dimension-0 part of this example (as in Figure 20.2(c)), we see that the rhombicuboctahedron has just one kind of vertex, which is only $\frac{1}{4}$-regular, having the symmetry of an isosceles trapezoid. Figure 20.3 shows the Schläfli symbols that correspond to quadrilaterals of varying degrees of regularity.

And naturally, we can read off the symbol of the dual of a polyhedron just by reversing its own symbol. For example, a rectangle and a rhombus are dual, and their symbols are reflected.

More Precise Definitions

More precisely, this new kind of Schläfli symbol is defined not really for an abstract polyhedron but for a polyhedron together with some group of its symmetries. This has already happened in our example: the half-regular embedded squares are geometrically regular squares, but the way they are embedded in the rhombicuboctahedron has only rectangular symmetry, since their sides alternately border triangles and squares.

Let's look at the Schläfli symbol of this kind of face, or equivalently, of a rectangle. How many types of edge does it have? The answer is two, since covering up the dimension-1 part leaves just two dots. Since a dot lies in just one row, we can see that each edge is regular; that is to say it has the symmetry interchanging its two ends. Covering up the dimension-0 part instead, we see that the Schläfli symbol of the vertex figure is a vertical line. This shows that there is only one kind of vertex, which is half regular, since this symbol has two rows, and indeed a rectangle does not have the symmetry that interchanges the two sides at a vertex.

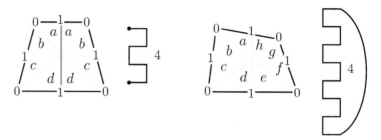

Verify that the isosceles trapezoid (above, left) has the three types of edges given by the dimension-1 coverup rule, namely two regular and one semi-regular, but only two kinds of vertices, both semi-regular, as given by the dimension-0 coverup rule. An irregular quadrilateral (above, right), which is only $\frac{1}{8}$-regular, has four kinds of vertices and four kinds of edges.

More General Definitions

The new kind of symbol is really defined for tessellations in an arbitrary simply connected space[1] of any dimension n, under arbitrary subgroups of their automorphism groups. In particular, it works for all topologically spherical polytopes. We shall give some interesting four- and eight-dimensional examples in our last chapter.

The general symbol will have a column for each dimension from 0 to n, the columns for dimensions 1 to $n - 1$ customarily being doubled for clarity. Moreover, each component of the symbol (under this doubling convention) receives a number. We now describe the rule that determines these numbers. How is this number determined?

---

[1] The symbol cannot distinguish between a non-simply connected space and any of its topological quotients; on the other hand, the symbol is unambiguous if we specify that the space is simply connected. In essence we are describing the way the simplices of a barycentric subdivision of the orbifold fit together.

For each d, the relation of d-joining is a permutation of order 2 on the flagstones of the tessellation—let us call this permutation π_d. It is easy to see that if d and d' differ by at least 2, then $\pi_d \pi_{d'}$ also has order 2. However, $\pi_d \pi_{d+1}$ can have any order and arbitrary cycle shape. The numbers attached to the components that lie between columns d and $d+1$ are the lengths of the corresponding cycles of $\pi_d \pi_{d+1}$.

Interlude: Polygons and Polytopes

The Types of Symmetry of a Polygon

Everybody is familiar with the classification of triangles into scalene, isoceles, and equilateral. This is really a classification by symmetry, and if our triangle has a pattern on it there is a fourth type, gyrational, as indicated by the arrows in Figure 20.4.

Quadrilaterals come in eight symmetry types, shown in Figure 20.3, although once again the gyrational type only happens for patterned squares.

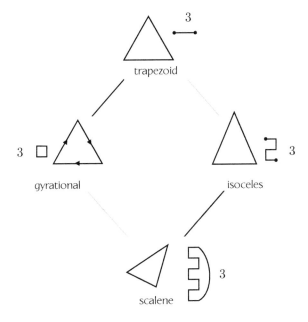

Figure 20.4. Types of triangles.

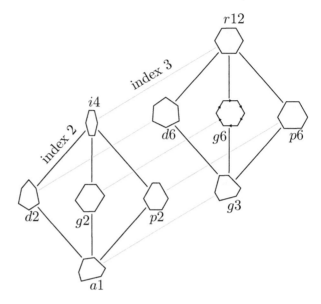

Figure 20.5. Types of hexagons.

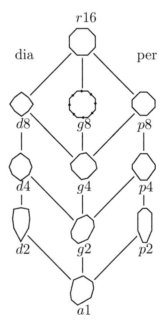

Figure 20.6. Types of octagons.

How many types of n-gons are there? The symmetry group of the regular n-gon is D_{2n}, the dihedral group of order $2n$, and what we are really doing here is drawing the lattice of subgroups of this, considered up to conjugacy. Each such subgroup has both an order and an index in D_{2n}, and the rule is that the number of conjugacy classes of groups of a given size is three if both order and index are even, and is one otherwise.

Figures 20.5 and 20.6 illustrate the types of hexagons and octagons and will help us to introduce some terminology. We'll call a polygon n-*symmetric* if its symmetry group has order n and $\frac{1}{r}$-*regular* to mean that its index is r.

Different types having the same order are distinguished by the nature of their symmetry axes. Such an axis may be a *diagonal*, meaning that it passes through two vertices, or a *perpend*, meaning that it is perpendicular to two edges. (In masonry, a "perpend stone" is one that abuts both sides of a wall.) We will call a polygon *diasymmetric* if all its axes are diagonal, *persymmetric* if they're all perpend, and *isosymmetric* if it either has both types or has axes that are both perpendicular to an edge and pass through the opposite vertex. (We may regard "isosymmetric" as abbreviating "having isosceles symmetry.") Finally, we call the symmetry of a polygon *gyrational* if it has no reflection axes, and we abbreviate "has gyrational symmetry" to "is gyrosymmetric."

This kind of terminology is also useful in higher dimensions, supplemented by "chirosymmetric" for "having chiral symmetry," since not all chiral symmetry groups are generated by gyrations.

Unfulfilled Groups

For unpatterned plane polygons, one kind of semiregularity $(= \frac{1}{2}$-regularity) does not appear, namely the gyrational type that corresponds to the cyclic subgroup C_n of D_{2n}. Any n-gon that has all the symmetries of C_n automatically has the remaining symmetries of D_{2n}. The same sort of thing happens for other categories of object, and we describe the missing subgroups as *unfulfilled* for the given category of object. For instance, for unpatterned polyhedra that are topologically cubes, the pyritohedral subgroup **3∗2** is unfulfilled.

Subgroup Lattices

We have some useful conventions for lattices of subgroups of small enough groups. A "card" represents a type of subgroup up to cojugacy, and the different styles of line represent containments to different prime indices (here thick for index 2, thin for index 3, and dotted, dashed or non-existent for higher indices, while composite indices are represented by combining these lines in the way shown in the figures, in which the gray spots may be regarded as "missing groups" (i.e., are unfilled).

We have found that this unorthodox convention handles complicated lattices so well that we earnestly recommend it to interested readers, along with the following extra conventions for the subgroup lattice of a group of form $2 \times \mathbf{G}$. Each rectangular card represents a pair of subgroups, of which the lower one $\mathbf{H}$ is a "pure subgroup" (i.e., a subgroup of $\mathbf{H}$) while the upper one $2 \times \mathbf{H}$ is its double. Any other (hexagonal) card lies on the edge between $\mathbf{H}$ and $\mathbf{K}$, where $\mathbf{H}$ (the Half group) has index 2 in $\mathbf{K}$, and denotes the third, "hybrid," subgroup "$\mathbf{HK}$" of $2 \times \mathbf{K}$ that contains $\mathbf{H}$.

If the generator of the direct summand group 2 is called -1, then $2 \times \mathbf{H}$ consists of the elements $+h$ and $-h$ for h in $\mathbf{H}$, while $\mathbf{HK}$ consists of the elements $+h$ and $-k$, for $h \in \mathbf{H}$, $k \in \mathbf{K} \backslash \mathbf{H}$.

Subgroups of Polyhedral Groups (Hendecacity)

Figures 20.7 and 20.8 show the lattices of subgroups of the polyhedral groups ∗**532** and ∗**432** under the above conventions, that for ∗**332** being visible in the latter. It is a useful mnemonic ("hendecacity") that the number of conjugacy classes of subgroups is 22 for ∗**532**, 33 for ∗**432**, and 11 for ∗**332**, always a multiple of 11.

Each group is given both an algebraic name, reflecting its structure as a pure or double or hybrid group, and a geometric name (its orbifold signature). In the algebraic names, a dihedral group of order $2n$ is called $\mathbf{D}_{2n}$ or $\mathbf{E}_{2n}$ according as it is generated by odd or even permutations; in a hybrid group $\mathbf{HK}$ we keep only the parameter of $\mathbf{K}$. In the cubic case there are several cases in which groups of the same signature are not conjugate in ∗**432**. When there are just two classes, we distinguish them by appending a $-$ sign if the group has odd (or diagonal) order-2 elements, or a $+$ sign when it has only even (or perpend) ones. In the ∗**22** case there are three signs, as shown below.

$$-\qquad\qquad \pm \qquad\qquad +$$

Note that the lower part of Figure 20.8 is the lattice of subgroups of $2 \times \mathbf{D}_8$, except that groups $\mathbf{C}_2$ and $\mathbf{E}_2$ have been identified, since they are conjugate in $\mathbf{S}[4]$.

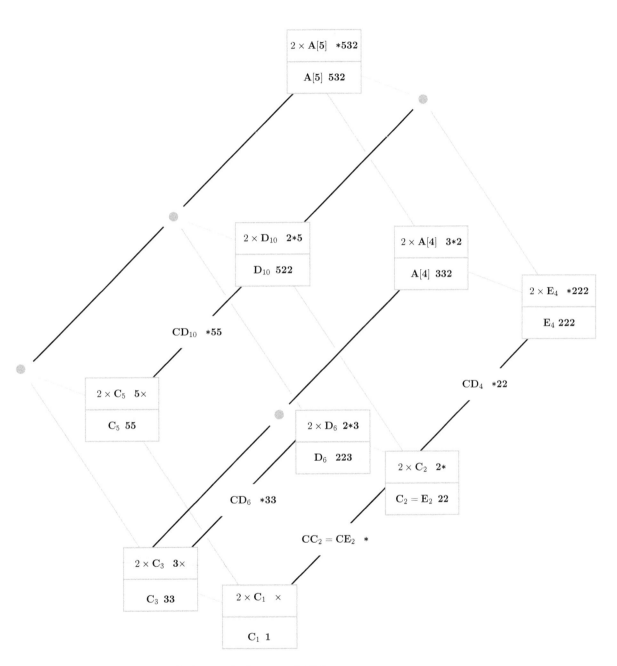

Figure 20.7. The lattice of subgroups of *532.

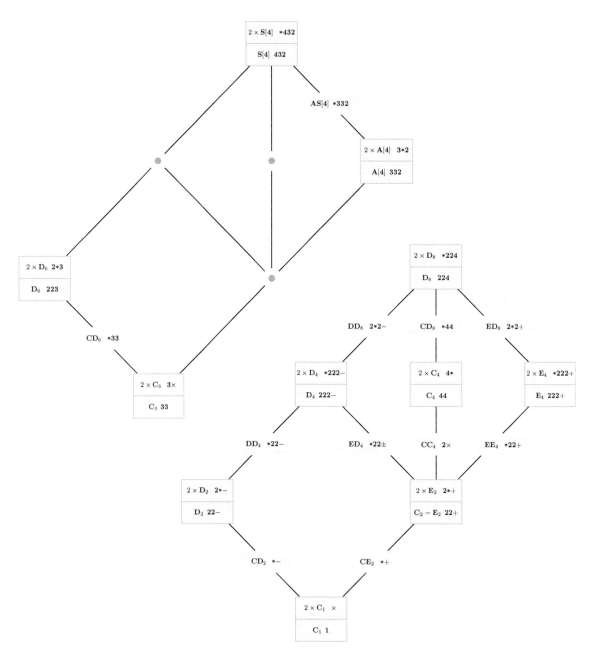

Figure 20.8. The lattice of subgroups of **∗432**. Each lower group is of index three in the corresponding group in the upper part of the figure.

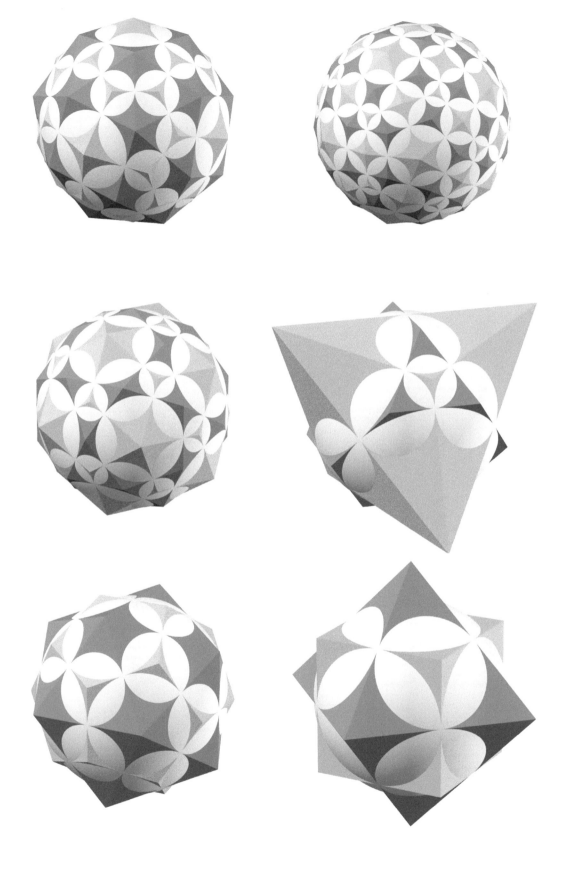

– 21 –

Naming Archimedean and Catalan Polyhedra and Tilings

The book in which Archimedes enumerated the polyhedra that have regular faces and equivalent vertices is unfortunately lost; however, its contents were reconstructed by Kepler, from whom the traditional names descend. In this chapter we explain these "Keplerian" names for the Archimedean and Catalan solids and extend them to the analogous tessellations of two- and three-dimensional Euclidean space.

We shall describe the polyhedra in dual pairs indicated by the arrows, and at the same time give our abbreviations for them, starting with the five Platonic (or regular) ones:

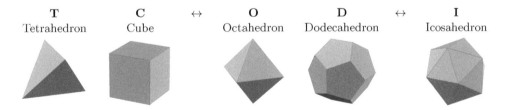

T	**C**	↔	**O**	**D**	↔	**I**
Tetrahedron	Cube		Octahedron	Dodecahedron		Icosahedron

Etymologically, the Greek stem "hedr-" is cognate with the Latin "sede-" and the English "seat," so that, for instance, "dodecahedron" really means "twelve-seater."

Truncation and "Kis"ing

These are followed by their "truncated" and "kis-" versions. Here, *truncation* means cutting off the corners in such a way that each regular n-gonal face is replaced by a regular $2n$-gonal one. The dual operation is to erect a pyramid on each face, thus replacing a regular m-gon by m isoceles triangles. These give five Archimedean and five Catalan solids:

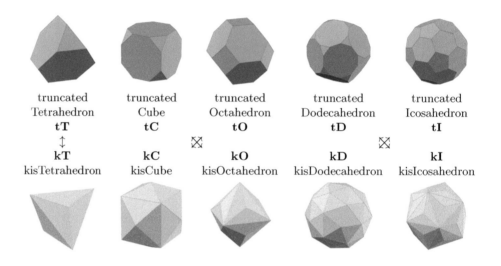

truncated	truncated	truncated	truncated	truncated
Tetrahedron	Cube	Octahedron	Dodecahedron	Icosahedron
tT	**tC**	**tO**	**tD**	**tI**
↕		⊠		⊠
kT	**kC**	**kO**	**kD**	**kI**
kisTetrahedron	kisCube	kisOctahedron	kisDodecahedron	kisIcosahedron

The names used by Kepler for the Catalan ones were rather longer, namely,

triakis	tetrakis	triakis	pentakis	triakis
tetrahedron	hexahedron	octahedron	dodecahedron	icosahedron

and were usually printed as single words. In these, "kis" meant "times," so that Kepler's "tetrakishexahedron" was literally a "(4×6)-seater." We retranslate "kis" as "multiplied," allowing us to abbreviate this to "kiscube," meaning "multiplied cube."[1]

---

[1]Beware! This is not the same as the "mucube" of later chapters.

Marriage and Children

We *marry* a regular polyhedron **P** and its dual **Q** by placing them so that corresponding edges intersect at right angles. Then, their *daughter* polyhedron is the Archimedean polyhedron that is their intersection, whose dual, their *son*, is the Catalan one that is their convex hull. This gives rise to only two new Archimedean and two new Catalan solids, which we can respectively truncate and "kis" to get two more of each:

CubOctahedron **CO**	IcosiDodecahedron **ID**	truncated CubOctahedron **tCO**	truncated IcosiDodecahedron **tID**
↕	↕	↕	↕
R₁₂ Rhombic dodecahedron	**R₃₀** Rhombic triacontahedron	**kR₁₂** kisRhombic dodecahedron	**kR₃₀** kisRhombic triacontahedron

The daughter and son of two mutually dual tetrahedra are the regular octahedron and cube, so that particular "marriage" leads to no new Archimedean and Catalan solids. However, we can incestuously marry the two Rhombic solids **R₁₂** and **R₃₀** to their respective duals **CO** and **ID**, producing two more Archimedean daughters and two more Catalan sons:

Polyhedral Sex

We call a polyhedron with V vertices and F faces *male* if $V > F$ (its virility exceeds its feminity) and *female* if $F > V$. Those with $V = F$ are hermaphrodites.

Trapezium and Trapezoid

If you look in a large enough dictionary you will probably find the assertions that "trapezium" (Br.) = "trapezoid" (U.S.) while "trapezoid" (Br.) = "trapezium" (U.S.)!

Although only the first of these is still true, it is interesting to see how this curious situation came about.

Proclus, a commentator on Euclid, used "trapezion" for a quadrilateral with (just) two parallel sides and "trapezoid" for a general four sided polygon with (typically) no parallel sides. All the European languages except English have maintained this usage. However, in English the words "trapezium" and "trapezoid" were accidentally interchanged in Hutton's *Mathematical Dictionary* of 1800 and this switched usage has persisted in the U.S.A. but was corrected in England between 1875 and 1900.

The end result has been that British "trapezium" and U.S. "trapezoid" have survived as synonyms, while British "trapezoid" and U.S. "trapezium" have been obsolete for more than a century, the word "quadrilateral" having been reintroduced to replace them.

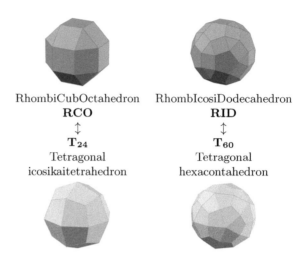

RhombiCubOctahedron
RCO
↕
T₂₄
Tetragonal
icosikaitetrahedron

RhombIcosiDodecahedron
RID
↕
T₆₀
Tetragonal
hexacontahedron

Strictly speaking, the solids we get by truncating **CO** and **ID** are not Archimedean since they have unequal edges, but they can be reformed to become so by distorting them suitably, and Kepler's names refer to the reformed versions. Their duals, the two Catalan solids we have described as *kisrhombic* have traditionally been called the hexakis octahedron and icosahedron. We prefer to use "hexakis" only for the replacement of a hexagon (rather than a triangle) by six triangles.

Since the traditional names here have often been misunderstood, we shall explain them. The prefix "rhombi-" in rhombicuboctahedron, for instance, does not refer to a supposed operation of "rhombi-truncation" but is an abbreviation for "rhombic dodecahedron," one of the two polyhedra **R₁₂** and **CO** of which **RCO** is the daughter.

In the names of the two Catalan solids, "icosi-kai-tetra" and "hexaconta" mean "twenty-plus-four" and "sixty," respectively, and "tetragonal" refers to the four-cornered-ness of the faces. Traditionally, the adjective has been "trapezoidal," but this is based on an obsolete meaning of the word "trapezoid" (see sidebar).

The two "rhombi" solids have some square faces that come from their father "rhombic" polyhedra, and we can divide each such square into two triangles in such a way as to increase the number of faces at each vertex from four to five and then reform the resulting two solids so that all their faces are regular.

Kepler's term for the first of the resulting two solids was "cubus simus" in which the second Latin word means "rounded" or "flattened." The "simian" apes are those with flattened noses. The traditional English names are snub cube (**sC**) and snub dodecahedron (**sD**), although they could equally be described as the snub octahedron and icosahedron. Their duals have respectively 24 and 60 pentagonal faces:

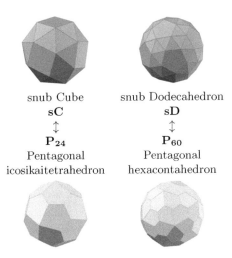

snub Cube	snub Dodecahedron
sC	**sD**
↕	↕
P₂₄	**P₆₀**
Pentagonal	Pentagonal
icosikaitetrahedron	hexacontahedron

Coxeter's Semi-Snub Operation

Coxeter pointed out that since the snub cube is equally the snub octahedron, it would be more sensible to regard it as being derived from the cuboctahedron, by a new type of snubbing operation. We call this *semi-snubbing*, abbreviated to "ssnub," because it is only half of the operation that leads from the cube to its ordinary snub: $\mathbf{C} \rightarrow \mathbf{CO} \rightarrow \mathbf{ssCO} = \mathbf{sC}$.

Regarded as the semi-snub cuboctahedron, the snub cube is obtained by distorting its daughter rhombicuboctahedron by twisting the regularly embedded (regular) squares in one direction, say, clockwise, which automatically twists the triangles counterclockwise and turns the non-regularly embedded (half-regular) squares into skew quadrilaterals that can be filled with two triangles. However, this is only a topological description; the canonical snub cube is obtained by a deformation that makes all the faces regular. (If we continue this twisting, until the skew quadrilaterals have no width at all, we obtain the cuboctahedron.)

The semi-snub can be defined topologically for an even-valence polyhedron in a similar way, but its faces can all be made regular only for some very special polyhedra. A notable example is the semi-snub mucube of Chapter 23.

Euclidean Plane Tessellations

Not many of the corresponding tilings of the Euclidean plane have previously received names. However, *quadrille* (**Q**), which we can interpret as "quadrangular grille," has been used for the standard tiling by squares. Based on this we propose *deltille* (**Δ**) for that by equilateral triangles and *hextille* (**H**) for that by regular hexagons. In these, the termination "tille" may be regarded as an amalgam of "tile" and "grille." The following table gives the resulting "Keplerian" names for the "Archimedean" and "Catalan" tilings of the plane. We have encountered their figures already, in Table 19.1 on pages 262–263.

Symbol	Archimedean Tiling	Face Code	Catalan Tiling	Symbol
Q	quadrille	4444	quadrille	**Q**
Δ	deltille	333333	hextille	**H**
H	hextille	666	deltille	**Δ**
tQ	tr. quadrille	488	kisquadrille	**kQ**
tH	tr. hextille	3 12 12	kisdeltille	**kΔ**
HΔ	hexadeltille	3636	rhombille	$\mathbf{R}_\infty$
tHΔ	tr. hexadeltille	4 6 12	kisrhombille	$\mathbf{kR}_\infty$
RHΔ	rhombihexadeltille	3464	tetrille	$\mathbf{T}_\infty$
sQ	snub quadrille	43343	4-fold pentille	$\mathbf{^4P}_\infty$
isQ	isosnub quadrille	44333	iso(4-)pentille	$\mathbf{i^4P}_\infty$
sH	snub hextille	63333	6-fold pentille	$\mathbf{^6P}_\infty$

Additional Data

We pause for additional data on the Archimedean polyhedra and planar tilings.

Vertex Figures

To any vertex V of a polyhedron, we associate the *vertex figure*, which is customarily obtained by slicing the polyhedron by a plane suitably near to V and perpendicular to the line joining V to the center of the polyhedron. Usually, each edge of the vertex figure is marked with the number of sides of the corresponding face of the polyhedron. Thus, the vertex figures of the cuboctahedron (Figure 21.1, left) and icosidodecahedron are rectangles whose sides are numbered 4,3,4,3 and 5,3,5,3. The vertex figures of the rhombicuboctahedron and rhombicosidodecahedron (Figure 21.1, right) are trapezoids with sides numbered 4,4,4,3 and 5,4,3,4.

The notion is analogously defined for higher-dimensional polytopes (Chapter 26) and can also be used for tessellations, despite the fact that this definition no longer applies. For example, the vertex figure of hexadeltille is a similar rectangle to the one in Figure 21.1 on the left, but now numbered 6,3,6,3.

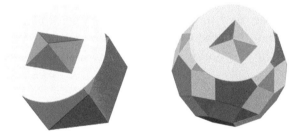

Figure 21.1. The vertex figures of the cuboactahedron and the rhombicosidodecahedron.

Wythoff Triangles

In a sense, the doily construction of Chapter 19 is a generalization of the *Wythoff triangle* (see Figure 21.2), a simple way to enumerate the tilings on which a kaleidoscopic group ∗**2pq** acts. The orbifold

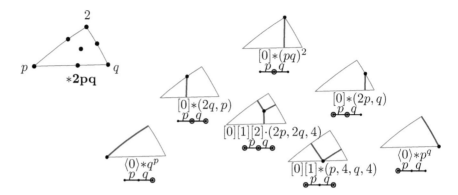

Figure 21.2. The Wythoff triangle; to each class of point in the triangle, there corresponds an archimedean tiling. The Wythoff-Coxeter symbol, discussed in Chapter 26 and shown beneath each triangle, indicates on which sides of the kaleidoscope the point lies.

for $*\mathbf{2pq}$ is of course a triangle, with vertices labeled $\mathbf{2}$, $\mathbf{p}$, and $\mathbf{qs}$. Metrically, the triangle has vertex angles $\frac{\pi}{2}$, $\frac{\pi}{p}$, and $\frac{\pi}{q}$, determining its geometry.

Topologically, there are just seven distinct classes of points on this orbifold: the three vertices, a point on one of the three edges, and an interior point. In each case, there is a canonical representative that is the vertex of an equilateral tiling by regular polygons.

More generally, as we will see in Chapter 26, Wythoff's construction can be used to generate higher-dimensional Archimedean tessellations, based on the Coxeter reflection groups. We postpone a discussion of the Wythoff-Coxeter symbol for these tessellations until that chapter.

The Data

In the following figures, we show data for the Archimedean polyhedra and planar tilings; we will use the dodecahedron and its descendants as examples (so, in the following figures $p = 5$ and $q = 3$). From left to right, we show the polyhedron; the Wythoff triangle, the archifold symbol, and the Wythoff-Coxeter symbol; the Schläfli symbol (Chapter 20); and the vertex figure. (The snub is chiral and does not arise from the Wythoff triangle, so does not receive a Wythoff-Coxeter symbol.)

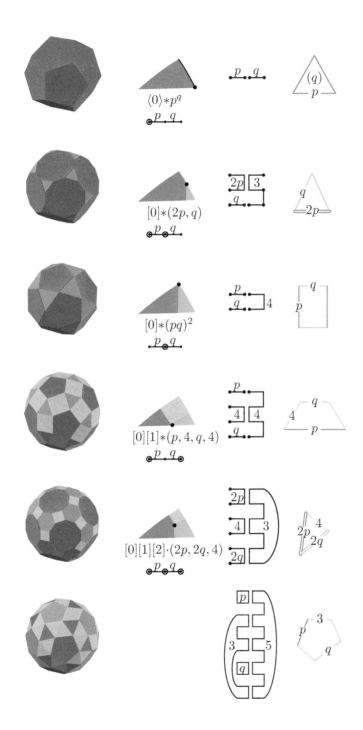

Architectonic and Catoptric 3-Tessellations

We shall use the term *Architectonic tessellation* for the three-dimensional analogs of the plane Archimedean tessellations of the previous section. The term is appropriate because Architectonics is the theory of structural design and because its beginning reminds us of Archimedes. The most interesting ones are those whose symmetry group is one of the "prime" space groups of Chapter 22, and we shall restrict ourselves to these.

Their duals turn out to be precisely those tessellations that can be obtained by repeated reflection of a suitable polyhedron in all its faces. Accordingly, we call them the *Catoptric tessellations*, since Catoptrics is the theory of mirror-reflections and because its beginning reminds us of Catalan. Once again, we restrict ourselves only to those with "prime" space groups.

There are thirteen tilings of each type, whose names are given in abbreviated form in Table 21.1 and more formally on the pages that follow it. The three rhombicuboctrilles are distinguished by the number of rhombicuboctahedra at each vertex and the three truncated cuboctrilles by their symmetry.

The simplest case is the tiling of space into cubes, which in analogy to quadrille we call *cubille*, for "cubic grille" (see Figure 21.3). This, being self-dual, belongs to both families and appears as the first line in Table 21.1, in which the tessellations are named by their most prominent cells.

Only four symmetry groups arise, which in the first column of the table we indicate as bc, nc, and fc for the groups $8°{:}2$, $4^-{:}2$, and

Figure 21.3. Cubille.

Symmetry Group	Flag Rank	Architectonic Name	Architectonic Cells	Catopric Name	
nc	1	cubille	C(8)	cubille	6c
fc	2	tetroctahedrille	T(8),O(6)	dodecahedrille	12d
d	4	trunctetrahedrille	tT(6),T(2)	obcubille	3d
bc	3	truncoctahedrille	TO(4)	obtetrahedrille	1/2d
nc	3	cuboctahedrille	CO(4),O(2)	oboctahedrille	4/2d
nc	4	tr. cubille	tC(4),O	pyramidille	1c
fc	8	tr. tetroctahedrille	tT(2),tO(2),CO	1/2-oboctahedrille	1d
nc	9	2-RCO-hedrille	RCO(2),CO,C(2)	1/4-oboctahedrille	2/4 e
fc	6	3-RCO-hedrille	RCO(3),C,T	1/4-cubille	3/2 e
nc	16	1-RCO-hedrille	RCO, tC,C,P8(2)	sq 1/4-pyramidille	1/4 c
nc	12	n-tCO-hedrille	tCO(2),tO,C	triangular ditto	1/4 e
fc	12	f-tCO-hedrille	tCO(2),tC,tO	1/2-pyramidille	1/2 e
bc	12	b-tCO-hedrille	tCO(2),P8(2)	1/8-pyramidille	1/2d

Table 21.1. Prime Architectonic and Catoptric tilings of space.

2^-:2 of the bicubic, normal cubic, and half-cubic lattices, respectively, and d for the diamond group 2^+:2. In the second, we give the flag rank, defined at the end of this chapter. The next two columns give the names and cells of the Architectonic tessellation (with their number at each vertex), while the penultimate one names the Catoptric chamber. Each of the catoptric chambers can be obtained by subdividing a cube (c), a rhombic dodecahedron (d), or either (e). We indicate which of these, and how many faces of the cube or rhombic dodecahedron are included in the chamber, in the final column of the table.

In the figures that follow, we show, starting on the far left and going clockwise, a portion of the architectonic tiling, a cell of the corresponding catoptric tiling shown relative to the architectonic cells, the Schläfli symbol of the architectonic tiling, a vertex figure, and, finally, a subdivision of the cube or rhombic dodecahedron or both into the catroptic cells. The cells of the architectonic tiling are colored red, green, blue, and yellow, corresponding to the code in Chapter 22. The symbols in the figure captions relate to those used in the table.

We comment briefly on the most interesting cases. *Tetroctahedrille* is the tessellation into alternating regular tetrahedra and octahedra. The catoptric that is dual to it is the one into rhombic dodecahedra. Both have appeared several times in our book.

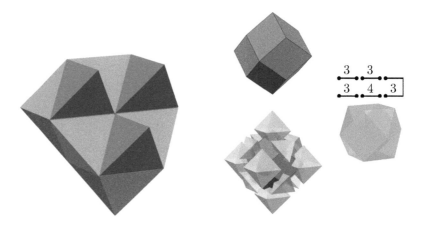

Tetroctahedrille. (*fc* 2) T(8),O(6) dodecahedrille 12*d*.

The chambers of the next three Catoptrics are "oblate" versions of the cube, tetrahedron, and octahedron, and so we call them *obcubille*, *oboctahedrille* and *obtetrahedrille*, respectively, more formally "oblate cubille," etc. Those of the later catoptrics are obtained by cutting these figures up—for instance, the oboctahedron decomposes into two square pyramids that are chambers for pyramidille, and these pyramids can be cut into two or four or eight in various ways to give chambers for the Catoptrics in the last few lines of Table 21.1.

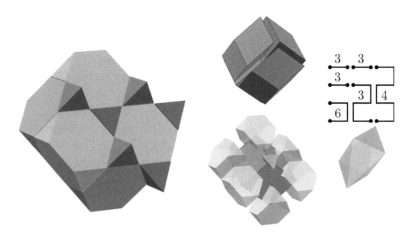

Trunctetrahedrille. (*d* 4) tT(6),T(2) obcubille 3*d*.

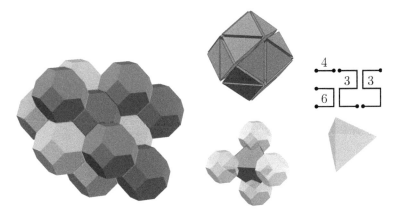

Truncoctahedrille. (*bc* 3) TO(4) obtetrahedrille 1/2*d*.

Cuboctahedrille. (*nc* 3) CO(4),O(2) oboctahedrille 4/2*d*.

The "trunc" in "truncoctahedrille" does not mean that this tiling is obtained by truncating another one. Rather, it indicates the fact that the cells of this tiling are truncated octahedra. However, "trunc cubille" and "trunc tetroctahedrille" are genuine truncations, the truncations of the Architectonic tilings of cubille and tetrahedrille, respectively. Note that truncations of Architectonic tilings are no longer Architectonic.

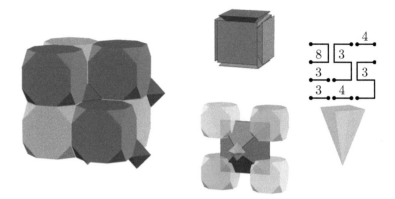

Trunc cubille. (*nc* 4) tC(4),O pyramidille 1*c*.

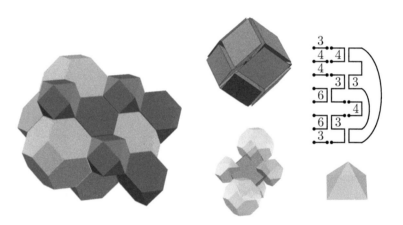

Trunc tetroctahedrille. (*fc* 8) tT(2),tO(2),CO 1/2–oboctahedrille 1*d*.

The three rhombicuboctahedrilles we call 1-fold, 2-fold, and 3-fold (1-RCO, 2-RCO, 3-RCO) according to the number of rhombicuboctahedra at each vertex, and (1 the chambers for their duals are quarters of various simpler ones. Perhaps the most interesting is *quarter cubille*, whose chambers are the quarter-cubes "subtended" at the center of a cube by the four faces of one of the inscribed tetrahedra—see the last figure on the next page.

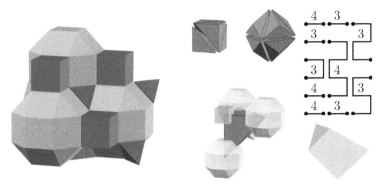

3-RCO-trille. (fc 6) RCO(3),C,T 1/4-cubille 3/2e.

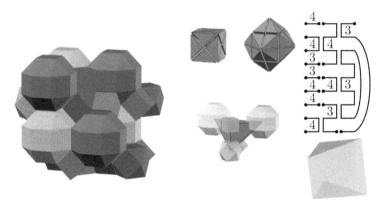

2-RCO-trille. (nc 9) RCO(2),CO,C(2) 1/4-oboctahedrille 2/4e.

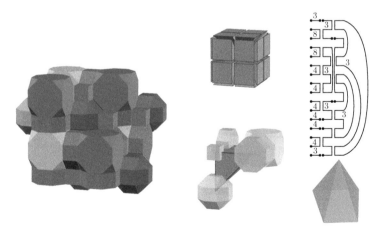

1-RCO-trille. (nc 16) RCO, tC,C,P8(2) square 1/4-pyramidille 1/4c.

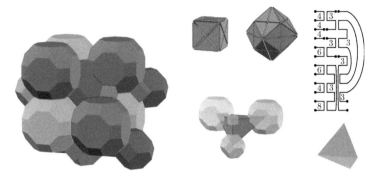

n-tCO-trille. (*nc* 12) tCO(2),tO,C triangular pyramidille 1/4*e*.

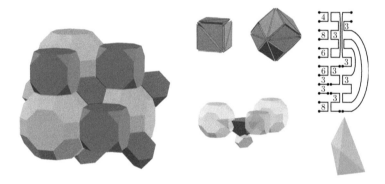

f-tCO-trille. (*fc* 12) tCO(2),tC,tO 1/2-pyramidille 1/2*e*.

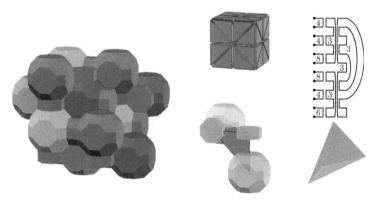

b-tCO-trille. (*bc* 12) tCO(2),P8(2) 1/8-pyramidille 1/8*c*.

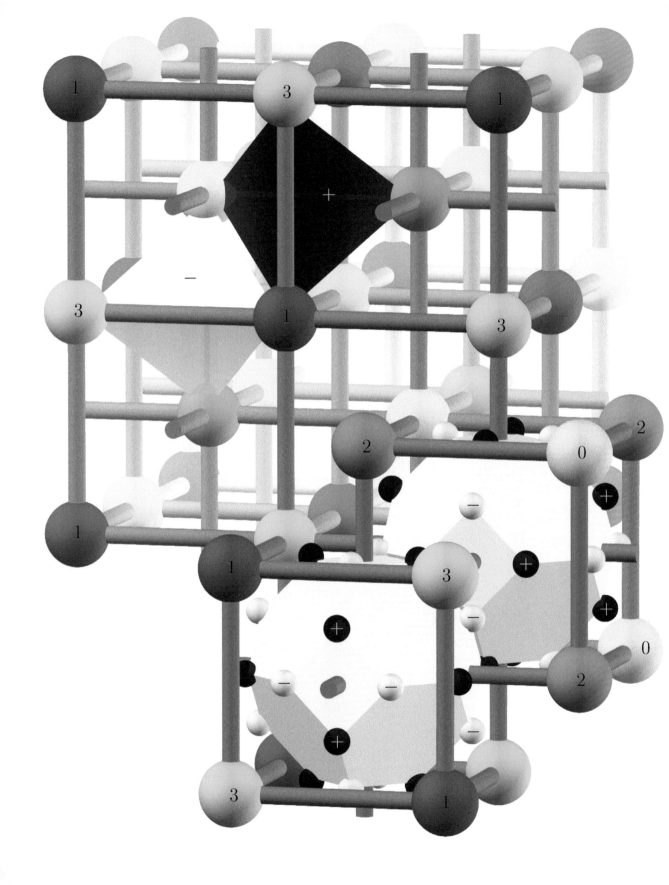

– 22 –
The 35 "Prime" Space Groups

In this chapter we discuss the 35 most interesting crystallographic space groups, namely the "prime" ones that don't fix any family of parallel lines. The less interesting ones that do fix such a family are naturally called "composite," since they can be obtained by compounding one- and two-dimensional groups; they are listed in Chapter 25.

This more mathematical chapter only describes the enumeration of prime groups—some things of which they describe the symmetries will appear in the next chapter.

The enumeration uses the particular families of points shown in the figure on the facing page, which we call *nodes*. Their coordinates (x, y, z) have the forms

$$
\begin{array}{ll}
0 & (\mathbb{Z}, \mathbb{Z}, \mathbb{Z}), x + y + z \text{ even} \\
2 & (\mathbb{Z}, \mathbb{Z}, \mathbb{Z}), x + y + z \text{ odd} \\
1 & (h, h, h), x + y + z - \frac{1}{2} \text{ even} \\
3 & (h, h, h), x + y + z - \frac{1}{2} \text{ odd} \\
+ & (\mathbb{Z}, q, h), (q, h, \mathbb{Z}) \text{ or } (h, \mathbb{Z}, q) \\
- & (\mathbb{Z}, q, h), (q, h, \mathbb{Z}) \text{ or } (h, \mathbb{Z}, q)
\end{array}
$$

where $\mathbb{Z}$ denotes any integer, h any integer $+\frac{1}{2}$, and q any integer $\pm\frac{1}{4}$.

In summary, the "digital" nodes $d = 0, 1, 2, 3$ are the (x, y, z) for which x, y, and z are *either* all integers *or* all integers $+\frac{1}{2}$, with $x + y + z = \frac{d}{2} \pmod 2$, while the "sign" nodes $+$ and $-$ are those with one coordinate of each type (integer, integer $\pm\frac{1}{4}$, integer $+\frac{1}{2}$), the sign being $+$ for that cyclic order and $-$ for the reverse.

Color Coding

In the pictures, we use:
yellow = 0
crimson = 1
blue = 2
green = 3
black = + = plus
white = – = minus

(opposite page) The 35 prime space groups can be understood by their action on the colored nodes shown here.

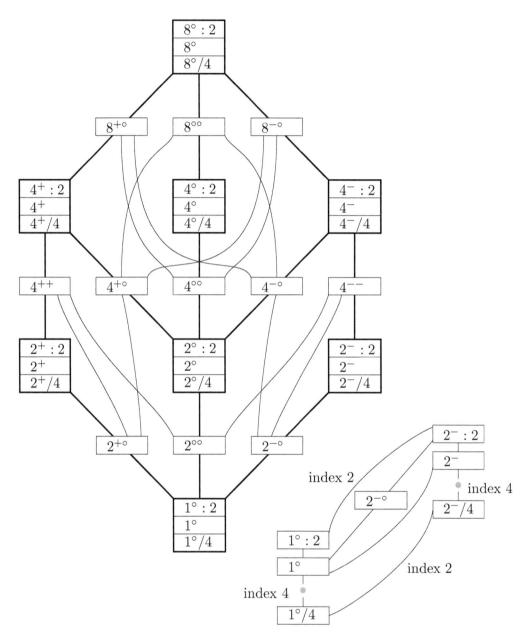

Figure 22.1. Relationships between the 35 prime space groups of this chapter. Bold edges represent multiple subgroup relations, as shown in the inset.

The Three Lattices

Collectively, the nodes 0, 1, 2, and 3 form two *normal cubic* (nc) lattices: 0 and 2 forming the usual one comprised of points with integer coordinates and 1 and 3 its translation by $(\frac{1}{2}, \frac{1}{2}, \frac{1}{2})$. Since each of these normal cubic lattices consists of the body-centers of the other, they together form what is usually called the *body-centered* (bc) cubic lattice—we prefer the shorter name *bicubic* lattice—fortunately, "bc" abbreviates both names.

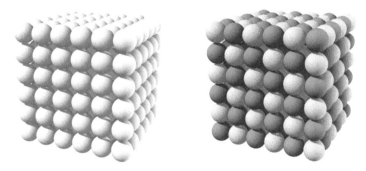

The normal cubic (nc) lattice, uncolored on the left and showing the colors of the nodes on the right.

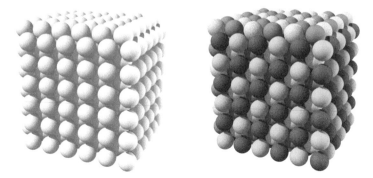

The bicubic (bc) lattice, again uncolored on the left and showing the colors of the nodes on the right.

The nodes numbered by any particular one of the four digits form what we call the *half-cubic* (fc) lattice since they are half the points of a cubic lattice. We have abbreviated "half" to its last letter "f" so that "fc" can also abbreviate the traditional name "face-centered cubic lattice." The new name makes it much easier to understand its symmetries, which are exactly half those of the normal cubic lattice.

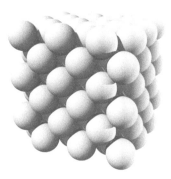

The half-cubic (fc) lattice, formed by the nodes of any one color.

The largest prime group **P** consists of all symmetries of the bc lattice. Each such symmetry effects some permutation of the symbols 0, 1, 2, 3, +, −, and 27 of the groups—the so-called *plenary groups*—are distinguished merely by saying which permutations they effect, as in the first part of the catalogue presented later in this chapter.

Displaying the Groups

We display the pure permutations (the ones that don't interchange + and −) inside a quartered box as follows:

I	(02)(13)		
		(0123)	(0321)
(13)	(02)		
		(01)(23)	(03)(21)

We also display their products with (+−) immediately above or below them:

I	$(02)(13)$	$(0123)(+-)$	$(0321)(+-)$
$(+-)$	$(02)(13)(+-)$	(0123)	(0321)
(13)	(02)	$(01)(23)(+-)$	$(03)(21)(+-)$
$(13)(+-)$	$(02)(+-)$	$(01)(23)$	$(03)(21)$

We use x for the *present* and o for the *absent* elements of any given group. For instance, the array for the group we call 4^-, whose permutations are $\{I, (02)(13), (02), (13)\}$, is

x	x	o	o
o	o	o	o
x	x	o	o
o	o	o	o

Translation Lattices and Point Groups

This arrangement makes it easy to pick out various things of interest, for instance, translation lattices and point groups. The subgroup formed by the translations in a given group is usually called its *translation lattice*: for the plenary groups it consists just of the translations of one of our three lattices, fc, nc, and bc. How do we tell which? Since the permutations effected by translations lie in the top row of our table, this is easily determined by the shape of the top row:

fc			
x	o	o	o

nc			
x	x	o	o

bc			
x	x	x	x

The *point group* of a given group is the finite group we get when we regard elements as identical if they differ merely by translations. This is also easily determined from the array, being **332** if the group is contained in the top row; **∗332**, **3∗2**, or **432** if it's confined to that row and the appropriate one of the other three rows; and finally **∗432** if it involves all four rows:

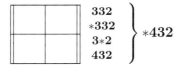

$$\left.\begin{array}{l}\textbf{332}\\ \textbf{∗332}\\ \textbf{3∗2}\\ \textbf{432}\end{array}\right\}\textbf{∗432}$$

Here are two useful ways of getting from one group to another. One can *enlarge the translation lattice* in the obvious way: to enlarge it all the way to bc one simply fills any nonempty row, while to go from fc to nc one replaces each x by an adjacent pair xx. Equally, one can *restrict the point group* into a possibly smaller one by discarding the parts that are not in the appropriate rows.

A *local subgroup* is one that fixes a point. The most interesting cases are those for which no line through the point is also fixed: i.e., the local subgroup is "prime." Since the point must be a node 0, 1, 2, or 3, these are also easily read from the array: the rule is that one intersects with

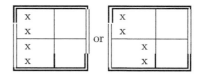

and then applies the point group rule.

Catalogue of Plenary Groups

For each group, the entry starts as follows:

Name of group, notation, international number

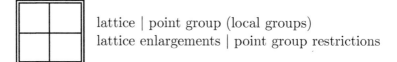

lattice | point group (local groups)
lattice enlargements | point group restrictions

Presentation by generators and relations

It first gives the array of group elements (x for present, o for absent), followed by a line giving the lattice (bc, nc, or fc), the point group (one of *432, 432, 3*2, *332, 332), and finally gives the lattice enlargements and point group restrictions (in the above orders). This is followed by generators and relations, then usually some examples of objects (to be described in the next chapter) whose symmetries are the given group, and possibly some further remarks.

The doubled octad group, $8°{:}2$, #229

$$bc \mid *\mathbf{432}(*\mathbf{432})$$
$$- \mid 8^{+°}, 8^{-°}, 4°{:}2, 4^{°°}$$

Presentation: $*\,\mathbf{4\ 6\ 2|4}$, meaning $*^{\mathbf{P}}\mathbf{4}^{\mathbf{Q}}\mathbf{6}^{\mathbf{R}}\mathbf{2}$ with $(PQRQ)^4 = 1$.

The largest plenary group $\mathbf{P}$, which is the symmetry group of the bc lattice and also the Scottish bubbles, the mucube, the muoctahedron, and the best lattice sphere-covering.

The pure octad group, $8°$, #223

$$nc \mid *\mathbf{432}\ (\mathbf{3}*\mathbf{2})$$
$$8°{:}2 \mid 4^{+}, 4^{-}, 4°, 2°$$

Presentation: $*\,\mathbf{4\ 4\ 3_2}$, meaning $*^{\mathbf{P}}\mathbf{4}^{\mathbf{Q}}\mathbf{4}^{\mathbf{R}}\mathbf{3}$ with $(PQRPRQ)^2 = 1$.

The symmetry group of tetrastix, also the Irish bubbles.

The negative hybrid octad group, $8^{-°}$, #204

$$bc \mid \mathbf{3}*\mathbf{2}\ (\mathbf{3}*\mathbf{2})$$
$$- \mid 4^{°°}$$

Presentation: $\mathbf{6}*\mathbf{2_2}$, meaning $^{\alpha}\mathbf{6}*^{\mathbf{P}}\mathbf{2}$ with $[\alpha^2, P]^2 = 1$.

The symmetry group of the icosahedral 3^9-hedron.

The null hybrid octad group, $8^{\circ\circ}$, #222

$$
\begin{array}{cc|cc}
\text{x} & \text{x} & \text{o} & \text{o} \\
\text{o} & \text{o} & \text{x} & \text{x} \\
\text{o} & \text{o} & \text{x} & \text{x} \\
\text{x} & \text{x} & \text{o} & \text{o}
\end{array}
$$

nc $|$ $*\mathbf{432}$ $(\mathbf{432})$
$8^{\circ}{:}4^{\circ-}, 4^{\circ+}, 4^{\circ}, 2^{\circ}$

Presentation: $\mathbf{4\ 6\ 2|4}$, meaning $^{\alpha}\mathbf{4}^{\beta}\mathbf{6}^{\gamma}\mathbf{2}$ with $(\alpha^{-1}\beta)^4 = 1$.

$(\mu C)_2 = (\mu O)_2$: the 2-chiral symmetries (i.e., those that fix the surface orientation) of the mucube or muoctahedron.

The positive hybrid octad group, $8^{+\circ}$, #211

$$
\begin{array}{cc|cc}
\text{x} & \text{x} & \text{x} & \text{x} \\
\text{o} & \text{o} & \text{o} & \text{o} \\
\text{o} & \text{o} & \text{o} & \text{o} \\
\text{x} & \text{x} & \text{x} & \text{x}
\end{array}
$$

bc $|$ $\mathbf{432}$ $(\mathbf{432})$
$-|4^{\circ\circ}$

Presentation: $*\ \infty\ \infty\ \mathbf{2_2|3}$, meaning $*^{\mathbf{P}}\infty^{\mathbf{Q}}\infty^{\mathbf{R}}\mathbf{2}$ with $(PQRPRQ)^2 = (PQRQ)^3 = 1$.

$(\mu C)_3 = (\mu O)_3$: the 3-chiral symmetries of the mucube and muoctahedron—since these form the chiral part of $8^{\circ}{:}2$, equally the chiral symmetries of the objects mentioned there. It is also the full symmetry group of tristix.

The negative doubled tetrad group, $4^{-}{:}2$, #221

$$
\begin{array}{cc|cc}
\text{x} & \text{x} & \text{o} & \text{o} \\
\text{x} & \text{x} & \text{o} & \text{o} \\
\text{x} & \text{x} & \text{o} & \text{o} \\
\text{x} & \text{x} & \text{o} & \text{o}
\end{array}
$$

nc $|$ $*\mathbf{432}$ $(*\mathbf{432})$
$8^{\circ}{:}2 \mid 4^{\circ-}, 4^{-}, 2^{\circ}{:}2, 2^{\circ}$

Presentation: $[4, 3, 4]$, meaning $*^{\mathbf{P}}\mathbf{4}^{\mathbf{Q}}\mathbf{3}^{\mathbf{R}}\mathbf{r}^{\mathbf{S}}\mathbf{2}$ with $(PR)^2 = (QS)^2 = 1$.

The symmetry group of the normal cubic lattice or tessellation.

The null doubled tetrad group, $4°{:}2$, #217

X	X	X	X
X	X	X	X
O	O	O	O
O	O	O	O

bc | $*\mathbf{332}$ ($*\mathbf{332}$)
$-$ | $4^{\circ\circ}$

Presentation: $\mathbf{4} * \mathbf{3_2}$, meaning $^{\alpha}\mathbf{4}*^{\mathbf{P}}\mathbf{3}$ with $[\alpha^2, P]^2 = 1$.

$ss\mu C$: the symmetries of the "semi-snub mucube" (i.e., the semi-snub of the multiplied cube).

The positive doubled tetrad group, $4^{+}{:}2$, #224

X	X	O	O
X	X	O	O
O	O	X	X
O	O	X	X

nc | $*\mathbf{332}$ ($*\mathbf{332}$)
$8°{:}2$ | $4^{+}{:}2, 4^{\circ+}, 2°{:}2, 2°$

Presentation: $* \mathbf{6}\ \mathbf{4}\ \mathbf{2_3}$, meaning $*^{\mathbf{P}}\mathbf{6}^{\mathbf{Q}}\mathbf{4}^{\mathbf{R}}\mathbf{2}$ with $(PQRPRQ)^3 = 1$ or $* \mathbf{6}\ \mathbf{6}\ \mathbf{2_2}$, meaning $*^{\mathbf{P}}\mathbf{6}^{\mathbf{Q}}\mathbf{6}^{\mathbf{R}}\mathbf{2}$ with $(PQRPRQ)^2 = 1$.

μCO : the symmetries of the mucuboctahedron.

The negative tetrad group, 4^{-}, #200

X	X	O	O
O	O	O	O
X	X	O	O
O	O	O	O

nc | $\mathbf{3}*\mathbf{2}$ ($\mathbf{3}*\mathbf{2}$)
$8^{-\circ}$ | $2°$

Presentation: $\bullet \overset{4/2}{\underset{}{\text{—}}}\overset{}{\textcircled{3}}\overset{4/2}{\underset{}{\text{—}}}\bullet$, meaning $1 = P^2 = [P, Q]^2 = Q^3 = [Q, R]^2 = R^2 = (RP)^2$.

The null tetrad group, $4°$, #218

x	x	o	o
o	o	x	x
o	o	o	o
o	o	o	o

nc | $\ast\mathbf{332}$ $(\mathbf{332})$
$4°{:}2$ | $2°$

Presentation: $\mathbf{4\ 4\ 3_2}$, meaning ${}^{\alpha}\mathbf{4}^{\beta}\mathbf{4}^{\gamma}\mathbf{3}$ with $(\alpha^2\beta^2)^2 = 1$.

$(ss\mu C)_2$: the 2-chiral symmetries of the semi-snub mucube.

The positive tetrad group, 4^+, #208

x	x	o	o
o	o	o	o
o	o	o	o
o	o	x	x

nc | $\mathbf{432}$ $(\mathbf{332})$
$8^{+°}|2°$

Presentation: $(\ast\ \mathbf{3\ 2\ 3\ 2})_2$, meaning $\ast^{\mathbf{P}}\mathbf{3}^{\mathbf{Q}}\mathbf{2}^{\mathbf{R}}\mathbf{3}^{\mathbf{S}}\mathbf{2}$ with $(PQRS)^2 = 1$.

$(\mu CO)_3$: the 3-chiral symmetries of the mucuboctahedron—equally the chiral symmetries of tetrastix or the Irish bubbles.

The binegative (hybrid) tetrad group, 4^{--}, #226

x	o	o	o		x	o	o	o
o	x	o	o	or	o	x	o	o
x	o	o	o		o	x	o	o
o	x	o	o		x	o	o	o

fc | $\ast\mathbf{432}$ $(\mathbf{3\ast 2,\ 432})$
$8°, 4^-{:}2$ | $2^{°-}, 2^-, 2^{°°}, 1°$

Presentation: $\overset{3}{\underset{}{③}}\!\!-\!\!\bullet\overset{4}{-}\!\!\bullet$, meaning $1 = P^3 = (PQ)^3 = Q^2 = (QR)^4 = R^2 = [P,R]$.

The symmetries of the snub-cubical 3^8-hedron. This group is the unique one that is "semisplit" in two distinct ways.

The negative hybrid tetrad group, $4^{\circ-}$, #207

x	x	o	o
o	o	o	o
o	o	o	o
x	x	o	o

nc $\mid$ **432** (**432**)

$8^{+\circ} \mid 2^{\circ}$

Presentation: $[4,3,4]^{+}$, meaning $^{\alpha}\mathbf{4}^{\beta}\mathbf{3}^{\gamma}\mathbf{4}^{\delta}\mathbf{2}$ with $(\alpha\gamma)^{2} = (\beta\delta)^{2} = 1$.

The chiral symmetries of the normal cubic lattice or tessellation.

The null hybrid tetrad group, $4^{\circ\circ}$, #197

x	x	x	x
o	o	o	o
o	o	o	o
o	o	o	o

bc $\mid$ **332** (**332**)

$-\mid-$

Presentation: $\infty \infty \, \mathbf{2_2}\vert\mathbf{3}$, meaning $^{\alpha}\infty^{\beta}\infty^{\gamma}\mathbf{2}$ with $(\alpha^{2}\beta^{2})^{2} = (\alpha^{-1}\beta)^{3} = 1$.

$(ss\mu C)_3$: the (3-)chiral symmetries of the semi-snub mucube and the snub-cubical 3^{8}-hedron. Also the symmetries of the propeller-hedron 3^{7}; this leads to the alternative presentation $^{\alpha}\mathbf{3}^{\beta}\mathbf{2}^{\mathbf{Z}}\times$ with $(\alpha Z\beta Z)^{2} = 1$. The appearance of $\times$ in this signature is connected with the fact that this is the unique prime space group not generated by its elements of finite order.

The positive hybrid tetrad group, $4^{\circ+}$, #201

x	x	o	o
o	o	o	o
o	o	x	x
o	o	o	o

nc $\mid$ **3∗2** (**332**)

$8^{-\circ}\vert 2^{\circ}$

Presentation: **6 6 $\mathbf{2_2}$**, meaning $^{\alpha}\mathbf{6}^{\beta}\mathbf{6}^{\gamma}\mathbf{2}$ with $(\alpha^{2}\beta^{2})^{3} = 1$.

The bipositive (hybrid) tetrad group, 4^{++}, #228

X	O	O	O
O	X	O	O
O	O	X	O
O	O	O	X

or

X	O	O	O
O	X	O	O
O	O	O	X
O	O	X	O

fc | $*\mathbf{432}$ $(\mathbf{332})$
$8^{-\circ}{:}2, 4^{+}{:}2 \mid 2^{+}, 2^{\circ+}, 2^{\circ\circ}, 1^{\circ}$

Presentation: $\mathbf{6\ 4\ 2_3}$, meaning $^{\alpha}\mathbf{6}^{\beta}\mathbf{4}^{\gamma}\mathbf{2}$ with $(\alpha^2\beta^2)^3 = 1$.

$(\mu CO)_2$: the 2-chiral symmetries of the mucuboctahedron. This is the unique plenary group that is neither split nor semisplit.

The negative doubled dyad group, $2^{-}{:}2$, #225

X	O	O	O
X	O	O	O
X	O	O	O
X	O	O	O

fc | $*\mathbf{432}$ $(*\mathbf{432}, \mathbf{3}*\mathbf{2})$
$8^{\circ}{:}2, 4^{-}{:}2 \mid 2^{\circ-}, 2^{-}, 1^{\circ}{:}2, 1^{\circ}$

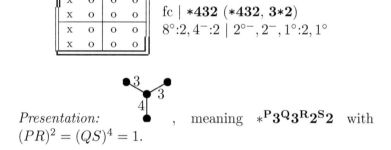

Presentation: , meaning $*^{\mathbf{P}}\mathbf{3}^{\mathbf{Q}}\mathbf{3}^{\mathbf{R}}\mathbf{2}^{\mathbf{S}}\mathbf{2}$ with
$(PR)^2 = (QS)^4 = 1$.

The symmetries of the fc lattice, the best lattice sphere-packing, and a salt crystal.

The null doubled dyad group, $2^{\circ}{:}2$, #215

X	X	O	O
X	X	O	O
O	O	O	O
O	O	O	O

nc | $*\mathbf{332}$ $(*\mathbf{332})$
$4^{\circ}{:}2 \mid 2^{\circ}$

Presentation: $*\ \mathbf{4\ 4\ 3}|\mathbf{3}$, meaning $*^{\mathbf{P}}\mathbf{4}^{\mathbf{Q}}\mathbf{4}^{\mathbf{R}}\mathbf{3}$ with
$(PQRQ)^3 = 1$.

$(ss\mu C)_1$: the 1-chiral symmetries of the semi-snub mucube.

The positive doubled dyad group, $2^+{:}2$, #227

X	O	O	O
X	O	O	O
O	O	X	O
O	O	X	O

or

X	O	O	O
X	O	O	O
O	O	O	X
O	O	O	X

fc | $*432$ ($*332$)
$8°{:}2, 4^+{:}2 \mid 2^+, 2^{\circ+}, 1°{:}2, 1°$

Presentation: $* \mathbf{6\ 6\ 2|3}$,   meaning $*^{\mathbf{P}}\mathbf{6}^{\mathbf{Q}}\mathbf{6}^{\mathbf{R}}\mathbf{2}$ with
$(PQRQ)^3 = 1$.

(μT): the symmetries of the mutetrahedron 6^6; equally those of a diamond crystal.

The negative dyad group, 2^-, #202

X	O	O	O
O	O	O	O
X	O	O	O
O	O	O	O

or

X	O	O	O
O	O	O	O
O	X	O	O
O	O	O	O

fc | $\mathbf{3*2}$ ($\mathbf{3*2}$, $\mathbf{332}$)
$8^{-\circ}, 4^- \mid 1°$

Presentation: $\bullet \overset{3}{\underset{}{\bullet}} \textcircled{3} \overset{4/2}{\underline{\quad}} \bullet$, meaning $1 = P^2 = (PQ)^3 = Q^3 = [Q, R]^2 = Q^2 = [P, R]$.

The symmetries of the dodecahedral packing.

The null dyad group, $2°$, #195

X	X	O	O
O	O	O	O
O	O	O	O
O	O	O	O

nc | $\mathbf{332}$ ($\mathbf{332}$)
$4^{\circ\circ} \mid 1°$

Presentation: $\mathbf{(3\ 2\ 3\ 2)_2}$,   meaning $^{\alpha}\mathbf{3}^{\beta}\mathbf{2}^{\gamma}\mathbf{3}^{\delta}\mathbf{2}$ with
$(\alpha\gamma)^2 = (\beta\delta)^2 = 1$.

$(ss\mu C)_{123}$: the symmetries of the semi-snub mucube that preserve all three orientations.

The positive dyad group, 2^+, #210

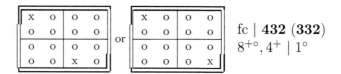

fc | **432 (332)**
$8^{+\circ}, 4^+ \mid 1^\circ$

Presentation: $(* \infty \infty \mathbf{3}|\mathbf{3})_\mathbf{3}$, meaning $*^\mathbf{P}\infty^\mathbf{Q}\infty^\mathbf{R}\mathbf{3}$ with $(PQRQ)^3 = (PQR)^3 = 1$.

$(\mu T)_3$: the (3-)chiral symmetries of either the mutetrahedron or a diamond crystal.

The negative hybrid dyad group, $2^{\circ-}$, #209

fc | **432 (432, 332)**
$8^{+\circ}, 4^{+\circ} \mid 1^\circ$

Presentation: $\overset{3}{\bullet\!-\!\!\underset{}{\overset{}{4}}\!\!\overset{3}{-\!\bullet}}$, meaning $1 = P^2 = (PQ)^3 = Q^4 = (QR)^3 = R^2 = [P, R]$.

The chiral symmetries of either the fc lattice or a salt crystal.

The null hybrid dyad group, $2^{\circ\circ}$, #219

fc | **∗332 (332)**
$4^\circ{:}2, 2^\circ{:}2|1^\circ$

Presentation: **4 4 3|3**, meaning $^\alpha\mathbf{4}^\beta\mathbf{4}^\gamma\mathbf{3}$ with $(\alpha^{-1}\beta)^3 = 1$.

The positive hybrid dyad group, $2^{\circ +}$, #203

X	O	O	O
O	O	O	O
O	O	X	O
O	O	O	O

or

X	O	O	O
O	O	O	O
O	O	O	X
O	O	O	O

fc | **3∗2 (332)**
$8^{-\circ}, 4^{\circ +} \mid 1^{\circ}$

Presentation: **6 6 2|3**, meaning $^{\alpha}\mathbf{6}^{\beta}\mathbf{6}^{\gamma}\mathbf{2}$ with $(\alpha^{-1}\beta)^3 = 1$.

$(\mu T)_2$: the 2-chiral symmetries of the mutetrahedron.

The doubled monad group, $1^{\circ}{:}2$, #216

X	O	O	O
X	O	O	O
O	O	O	O
O	O	O	O

fc | **∗332 (∗332)**
$4^{\circ}{:}2, 2^{\circ}{:}2 \mid 1^{\circ}$

Presentation: $\square$, meaning $\ast^{\mathbf{P}}\mathbf{3}^{\mathbf{Q}}\mathbf{3}^{\mathbf{R}}\mathbf{3}^{\mathbf{S}}\mathbf{3}$ with $(PR)^2 = (QS)^2 = 1$.

$(\mu T)_1$: the 1-chiral symmetries of the mutetrahedron.

The monad group, 1°, #196

X	O	O	O
O	O	O	O
O	O	O	O
O	O	O	O

fc | **332 (332)**
$4^{\circ\circ}, 2^{\circ} \mid -$

Presentation: $\square^{+}$, meaning $^{\alpha}\mathbf{3}^{\beta}\mathbf{3}^{\gamma}\mathbf{3}^{\delta}\mathbf{3}$ with $(\alpha\gamma)^2 = (\beta\delta)^2 = 1$.

$(\mu T)_{123}$: the symmetries of the mutetrahedron that preserve all three orientations. This is the smallest plenary group, **p**.

The Quarter Groups

In the plenary groups there are four axes of 3-fold rotations ("triad axes") through each node 0, 1, 2, 3. The quarter groups retain just one of these, in a subtle way.

The eight pure groups

$$8°, \; 4^-, \; 4°, \; 4^+, \; 2^-, \; 2°, \; 2^+, \; 1°$$

have subgroups of index 4 that we naturally call the *quarter groups*:

$$8°/4, \; 4^-/4, \; 4°/4, \; 4^+/4, \; 2^-/4, \; 2°/4, \; 2^+/4, \; 1°/4,$$

defined by fixing a distinguished axis through each digit or node. These are described more fully in the next chapter.

The rule is that the direction of the distinguished axis through a given node is

$$\pm(+,+,+) \quad \pm(+,-,-) \quad \pm(-,+,-) \quad \pm(-,-,+)$$

if the coordinates of the node are respectively congruent (mod 2) to

$$(0, 0, 0) \qquad (1, 1, 0) \qquad (0, 1, 1) \qquad (1, 0, 1)$$
$$(1, 1, 1) \qquad (0, 0, 1) \qquad (1, 0, 0) \qquad (0, 1, 0)$$
$$\left(\tfrac{1}{2},\tfrac{1}{2},\tfrac{1}{2}\right) \quad \left(-\tfrac{1}{2},\tfrac{1}{2},-\tfrac{1}{2}\right) \quad \left(-\tfrac{1}{2},-\tfrac{1}{2},\tfrac{1}{2}\right) \quad \left(\tfrac{1}{2},-\tfrac{1}{2},-\tfrac{1}{2}\right)$$
$$\left(-\tfrac{1}{2},-\tfrac{1}{2},-\tfrac{1}{2}\right) \quad \left(\tfrac{1}{2},-\tfrac{1}{2},\tfrac{1}{2}\right) \quad \left(\tfrac{1}{2},\tfrac{1}{2},-\tfrac{1}{2}\right) \quad \left(-\tfrac{1}{2},\tfrac{1}{2},\tfrac{1}{2}\right)$$

The rule is "Integers Advance, Fractions Retard," meaning that if the odd-coordinate-out (mod 2) for the point is x or y or z and the coordinates are integers, then the odd-coordinate-out for the direction is y or z or x, while if they are fractions, it is z or x or y. When there is no odd-coordinate-out for the point, then there is none for the direction.

Catalogue of Quarter Groups

This catalogue has the same form as that for the Plenary groups, except that the correspondence of translation lattices to top rows becomes

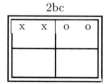

and there are no "prime" local groups. Some of the presentations involve bars, and others "fractional powers," which are explained in the appendix to this chapter.

The quarter octad group, $8°/4$, #230

x	x	o	o
o	o	x	x
x	x	o	o
o	o	x	x

2bc | *$\mathbf{432}$
$-|4^+/4, 4^-/4, 4°/4, 2°/4$

Presentation: $\overline{4}\,\overline{6}\,\mathbf{2}$, meaning $^\alpha\mathbf{4}^\beta\mathbf{6}^\gamma\mathbf{2}$ with $(\alpha\gamma\alpha^2\gamma\alpha^3\gamma)^2 = 1$ or $(\beta\gamma\beta^2\gamma\beta^3\gamma)^2 = 1$.

The largest quarter group, $\mathbf{Q}$. Best thought of as the (larger) symmetry group (**4stakes**) of tetrastakes (equally of checkerstix). It is also the symmetry group of hexastix (**6stix**) and the musnub cube (μs$\mathbf{C}$).

The negative quarter tetrad group, $4^-/4$, #206

x	x	o	o
o	o	o	o
x	x	o	o
o	o	o	o

2bc | $\mathbf{3*2}$
$-|2°/4$

Presentation: $\overline{6}\,\overline{6}\,\mathbf{2}$, meaning $^\alpha\mathbf{6}^\beta\mathbf{6}^\gamma\mathbf{2}$ with $(\alpha\beta^2\alpha^3\beta\alpha^2\beta^3)^2 = 1$.

(**6stix**)$_2$: the "2-chiral" symmetries of hexastix, that is, the chiral symmetries that fix directions of the "pencils" together with the achiral ones that reverse these directions.

The null quarter tetrad group, $4°/4$, #220

x	x	o	o
o	o	x	x
o	o	o	o
o	o	o	o

2bc | *$\mathbf{332}$
$-|2°/4$

Presentation: $\overline{4}\,\overline{4}\,\mathbf{3}$, meaning $^\alpha\mathbf{4}^\beta\mathbf{4}^\gamma\mathbf{3}$ with $(\alpha\beta^2\alpha^3\beta\alpha^2\beta^3)^2 = 1$.

6stakes: the symmetries of hexastakes.

The positive quarter tetrad group, $4^+/4$, #214

X	X	o	o
o	o	o	o
o	o	o	o
o	o	X	X

$2bc \mid \mathbf{432}$
$- \mid 2°/4$

Presentation:  , meaning $1 = P^2 =$ "$(PQ)^{10/3}$" $= Q^3 =$ "$(QR)^{10/2}$" $= R^2 = (PR)^2$.

(**4stakes**)$_3$ = (**6stix**)$_3$: this is the chiral part of $8°/4$, so it also consists of the chiral symmetries of the other objects mentioned there.

The negative quarter dyad group, $2^-/4$, #205

X	o	o	o
o	o	o	o
X	o	o	o
o	o	o	o

or

X	o	o	o
o	o	o	o
o	X	o	o
o	o	o	o

$2nc \mid \mathbf{3*2}$
$4^-/4 \mid 1°/4$

Presentation: $\bar{6}\ \bar{6}\ \mathbf{3}$, meaning $^\alpha\mathbf{6}^\beta\mathbf{6}^\gamma\mathbf{3}$ with $\alpha^2\beta = \beta^2\alpha$.

The symmetries of birhombohedrille, a tiling of space that alternates rhombohedra of two different shapes (left).

The null quarter dyad group, $2°/4$, #199

X	X	o	o
o	o	o	o
o	o	o	o
o	o	o	o

$2bc \mid \mathbf{332}$
$- \mid 1°/4$

Presentation: , meaning $1 = P^2 =$ "$(PQ)^{10/3}$" $= Q^3$.

(**6stakes**)$_3$: the (3-)chiral symmetries of hexastakes.

The positive quarter dyad group, $2^+/4$, #212, #213

X	O	O	O
O	O	O	O
O	O	O	O
O	O	X	O

or

X	O	O	O
O	O	O	O
O	O	O	O
O	O	O	X

2nc | **432**
$4^+/4 \mid 1°/4$

Presentation: $\overset{10/2}{③\!\!-\!\!\bullet}$, meaning $1 = Q^3 = \text{"}(QR)^{10/2\text{"}} = R^2$.

This the unique one of the 35 prime groups that is "metachiral," that is to say it is distinct from its mirror image group. The two enantiomorphous forms have international numbers 212 and 213.

The quarter monad group, $1°/4$, #198

X	O	O	O
O	O	O	O
O	O	O	O
O	O	O	O

2nc | **332**
$2°/4 \mid -$

Presentation: $\overset{5/2}{③\!\!-\!\!③}$, meaning $1 = P^3 = \text{"}(PQ)^{5/2\text{"}} = Q^3$.

$(\mathbf{4stakes})_{123}$: the symmetries of tetrastakes that fix all three orientations; equally, the color-preserving symmetries of the snub mucube. This is the smallest quarter group, **q**.

Why This List Is Complete

Our list of the 35 prime groups is taken, like the list of 184 composite ones in Chapter 25, from [6]. A simple proof is given there that the list is complete, by first using an algebraic argument to show that any such group must contain elements of order 3 and then using a geometric one to show that there are only two configurations for the axes of these. The group is a plenary group if these axes may intersect and otherwise a quarter group.

When the configuration of triad axes is known, the group must be contained in their symmetry group, which is **P** or **Q**, and must contain the group generated by the order-3 rotations around them, which is **p** or **q**. Since the later containment is obviously normal,

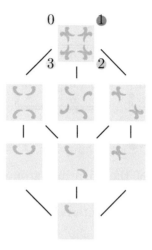

Figure 22.2. The enumeration of the prime groups reduces to that of the subgroups of $2 \times \mathbf{D_8} \cong \mathbf{P/p}$ and $\mathbf{D_8} \cong \mathbf{Q/q}$. $\mathbf{D_8}$ (as do its subgroups) acts on the symbols $0, 1, 2, 3$ as shown here, and $2 \times \mathbf{D_8}$ (etc.) acts on $+, -$ and $0, 1, 2, 3$.

this reduces the problem to enumerating the subgroups of the finite quotient groups $\mathbf{P/p} \cong 2 \times \mathbf{D_8}$ and $\mathbf{Q/q} \cong \mathbf{D_8}$. This enumeration is quite easy and is illustrated in Figure 22.2.

Appendix: Generators and Relations for the 35 Groups

The presentations by generators and relations that we have given are in many cases associated to infinite polyhedra such as the mucube (see the next chapter for its definition!). We illustrate the idea by describing how that for $8°{:}2$ may be derived from the mucube.

The universal cover of the mucube is a map of squares in the hyperbolic plane, six to a vertex, whose group is obviously $*\mathbf{462}$ with presentation

$$*^{\mathbf{P}}\mathbf{4}^{\mathbf{Q}}\mathbf{6}^{\mathbf{R}}\mathbf{2} : 1 = P^2 = (PQ)^4 = Q^2 = (QR)^6 = R^2 = (RP)^2.$$

In this (see Figure 22.3) P and Q are reflections fixing a square, while R reflects in an edge of the square. It follows that the conjugate $R^Q = QRQ$ of R by Q reflects in an edge not meeting P, and so the product $PR^Q = PQRQ$ translates the given square to an adjacent one. But, the fourth power of this translation maps to the identity in the mucube because of the way the four squares form a ring.

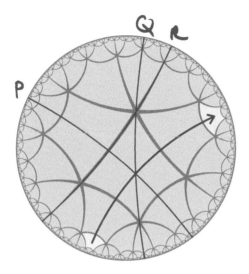

Figure 22.3. A presentation for *462.

Moreover, it is easy to see geometrically that this new *wrapping* relation $(PQRQ)^4 = 1$ determines the mucube and so completes the above presentation for *462 to one for the mucube's symmetry group $8^\circ{:}2$. (See Figure 22.4.)

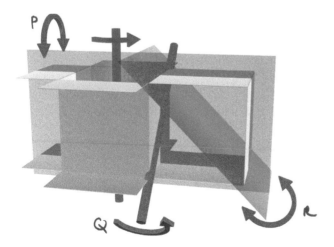

Figure 22.4. A presentation for $*^P4^Q6^R2$.

Naming the Wrapping Relations

We call the resulting presentation $* \mathbf{4}\ \mathbf{6}\ \mathbf{2}|\mathbf{4}$; this name extends the signature $*\mathbf{462}$ analogously to the way Coxeter's name $[4, 6 \mid 4]$ extends his name $[4, 6]$ for the hyperbolic group. It is appropriate since the final numbers 2 and 4 are alternatives: if we used $P, Q, R' = R^Q$ as generators, the presentation would become

$$*\mathbf{4}\ \mathbf{6}\ \mathbf{4}|\mathbf{2} : 1 = P^2 = (PQ)^4 = Q^2 = (QR')^6 = R'^2 = (R'P)^4$$

with the wrapping relation $(PQR'Q)^2 = 1$.

The subgroup Coxeter calls $[4, 6 \mid 4]^+$ generated by $\alpha = PQ, \beta = QR, \gamma = RP$ we analogously call $\mathbf{4}\ \mathbf{6}\ \mathbf{2}|\mathbf{4}$, meaning that its presentation is obtained from

$$^\alpha\mathbf{4}^\beta\mathbf{6}^\gamma\mathbf{2} : \alpha\beta\gamma = 1 = \alpha^6 = \beta^4 = \gamma^2$$

by adjoining the wrapping relation $(\alpha\beta^{-1})^4 = 1$. This "new" wrapping relation is really the same as the old one, since $\alpha\beta^{-1} = PQRQ$.

In a similar way, the subscript 2 that defines the wrapping relation in the related cases $\mathbf{4}\ \mathbf{4}\ \mathbf{3_2}$, $* \mathbf{4}\ \mathbf{4}\ \mathbf{3_2}$, and $\mathbf{4} * \mathbf{3_2}$ indicates the order of an element that is really the same in all three cases, although its expression in terms of their generators will vary. The new relation is $(\alpha\gamma^{-1}\beta)^2 = 1$ or $(\alpha^2\beta^2)^2 = 1$ for $^\alpha\mathbf{4}^\beta\mathbf{4}^\gamma\mathbf{3}$; equivalently, $(PQPRQR)^2 = 1$ or $(Q^P Q^R)^2 = 1$ for $*^\mathbf{P}\mathbf{4}^\mathbf{Q}\mathbf{4}^\mathbf{R}\mathbf{3}$, and $[\alpha^2, P]^2 = 1$ for $^\alpha\mathbf{4}*^\mathbf{P}\mathbf{3}$.

Finally, the subscript appended to the entire name in the cases $(*\mathbf{3232})_\mathbf{2}$ and $(\mathbf{3232})_\mathbf{2}$ indicates the symmetric relation $(PQRS)^2 = 1$ for $*^\mathbf{P}\mathbf{3}^\mathbf{Q}\mathbf{2}^\mathbf{R}\mathbf{3}^\mathbf{S}\mathbf{2}$, which becomes the pair of relations $(\alpha\gamma)^2 = 1 = (\beta\delta)^2$ in $^\alpha\mathbf{3}^\beta\mathbf{2}^\gamma\mathbf{3}^\delta\mathbf{2}$, since $\alpha\gamma = PQRS$ and $\beta\delta = QRSP$.

Coxeter-Type Presentations

The groups $4^-{:}2$, $2^-{:}2$, and $1°{:}2$ are generated by reflections, so Coxeter diagrams (treated more fully in Chapter 26) are appropriate for them. We shall also adopt Coxeter's convention of appending a + sign to indicate the "chiral subgroup." The nodes in these diagrams represent elements of order 2, while a branch marked n between nodes P and Q indicates the relation $(PQ)^n = 1$. Marks of 3 are usually omitted, while generators that correspond to unjoined nodes commute.

We extend these conventions in several ways. First, a generator of order $n > 2$ is represented by a circled n. Second, the "alternative" relation $(PQRQ)^n = 1$ is indicated by putting $\hat{n}$ near Q and between P and R, as in the marginal figure for **4 6 2|4**. The third extension deserves a section of its own.

Fractional Powers and the Chiral Quarter Groups

The "fractional power" notation is exemplified by

$$\text{``}(PQ)^{4/2}\text{''} = \bar{P}\bar{Q}PQ\bar{P}\bar{Q}PQ,$$
$$\text{``}(PQ)^{5/3}\text{''} = \bar{P}QP\bar{Q}\bar{P}Q\bar{P}\bar{Q}PQ,$$
$$\text{``}(PQ)^{10/3}\text{''} = \bar{P}\bar{Q}\bar{P}QPQ\bar{P}\bar{Q}\bar{P}\bar{Q}PQP\bar{Q}\bar{P}\bar{Q}PQPQ,$$
$$\text{``}(QR)^{10/2}\text{''} = \bar{Q}\bar{R}\bar{Q}\bar{R}\bar{Q}RQRQR\bar{Q}\bar{R}\bar{Q}\bar{R}\bar{Q}RQRQR,$$

in which the bars denote inverses. These appear in our presentations for the chiral quarter groups.

The general rule here is as follows. We obtain "$(PQ)^{n/d}$" by writing PQ n times and then alternately barring and unbarring stretches of average length n/d. For example, "$(PQ)^{2/1}$" denotes the commutator $\bar{P}\bar{Q}PQ$. When n/d is not an integer, the exact rule is as follows: write the multiples of n/d up to $2n$ as mixed fractions; for instance, in the case $10/3$ these are $0, 3\frac{1}{3}, 6\frac{2}{3}, 10, 13\frac{1}{3}, 16\frac{2}{3}, 20$. Then, the differences of their integer parts, namely $3, 3, 4, 3, 3, 4$, give the lengths of the barred and unbarred stretches.

These curious relations really come from knot theory; for example, the one for $1°/4$ corresponds to Thurston's wonderful observation that the orbifold of this group is a 3-sphere in which the figure eight knot is a "cone-line" of order 3. The figure eight knot is the rational knot $5/3$, whose fundamental group has the presentation

$$P, Q : \text{``}(PQ)^{5/3}\text{''} = 1.$$

Similarly, the orbifold of $2°/4$ is a 3-sphere containing the rational link $10/3$ in which the components are cone-lines of distinct orders 2 and 3.

Coxeter found presentations for some complex reflection groups that involve such fractional power relations.

The Achiral Quarter Groups

Each of the four remaining groups—$8^-/4$, $4^-/4$, $4°/4$, $2^-/4$—can be generated by two rotatory reflections of orders p and q, say, whose product is a rotation of order r. The group is therefore a quotient of **p q r**, which we call **p̄ q̄ r**.

   This symbol does not actually specify a presentation, since in each case there are infinitely many ways to select such generators. However, it appears that the simplest choices are essentially unique: they lead to the wrapping relations that were given in the text.

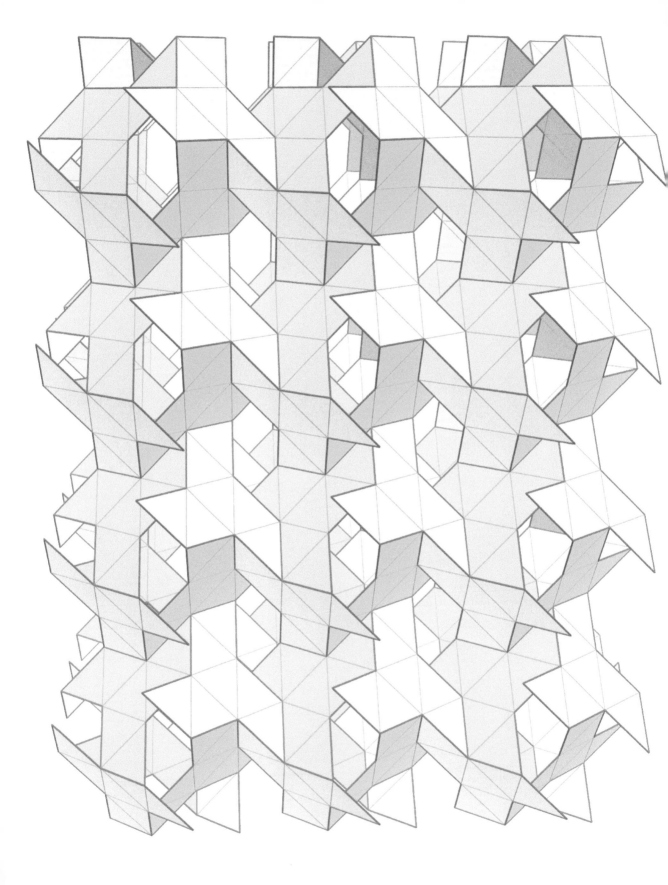

– 23 –

Objects with Prime Symmetry

The purpose of this chapter is to describe the objects used as examples in Chapter 22.

The Three Lattices

The symmetries of the three lattices are easily found. Since the bc lattice consists of all points $0, 1, 2, 3$ and doesn't distinguish $+$ from $-$, its symmetries can achieve the full dihedral group of permutations of $0, 1, 2, 3$ and also the interchange of $+$ and $-$. They therefore constitute the *doubled octad group* ($8°{:}2$). The nc lattice consists just of 0 and 2, whose symmetries effect $\langle (02), (13), (+-) \rangle$, determining the *negative double tetrad group* ($4^-{:}2$). Finally, the symmetries of the fc lattice, whose points are those colored 0, effect only $\langle (13), (+-) \rangle$, so form the *negative double dyad group* ($2^-{:}2$).

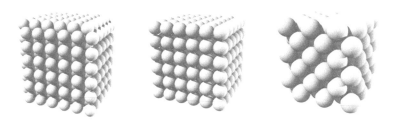

The three cubic lattices: The bc lattice has symmetry $8°{:}2$, the nc has $4^-{:}2$, and the fc has $2^-{:}2$.

(opposite page) The fascinating propeller-hedron has group $4°°$.

Each of the three lattices actually gives five groups, one for each point group:

point group	*432	432	3*2	*332	332
bc lattice	8°:2	8⁺°	8⁻°	4°:2	4°°
nc lattice	4⁻:2	4°⁻	4⁻	2°:2	2°
fc lattice	2⁻:2	2°⁻	2⁻	1°:2	1°

The additional groups in the table are obtained by restricting to the given point groups. Assemblies with these as symmetries are easily obtained by surrounding the lattice points by finite objects that force the appropriate point group and are all parallel to each other.

One such assembly gives the densest known packing of regular dodecahedra (see Figure 23.1): these dodecahedra are arranged much like the rhombic ones we will see shortly in Figure 23.4, so alternate cubes of the cubic lattice are inscribed in them. The packing has density $\rho = .9405$, the unused space consisting of curious polyhedra we call *endo*-dodecahedra, which like regular dodecahedra have 12 pentagonal faces with all edges of the same length although neither the faces nor the polyhedron are convex. Since the dodecahedra are centered on the fc lattice and force the pyritohedral group **3*2**, the symmetry group is 2^-.

Although the cubic tessellation has group 4^-:2, a model of it made with the Zome System construction set (a proprietary system

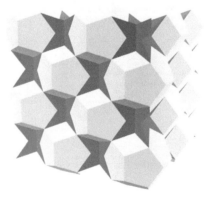

Figure 23.1. A packing of regular dodecahedra. This half-cubic lattice of dodecahedra has point group **3*2** and symmetry 2^-. The endo-dodecahedral holes are shown in orange.

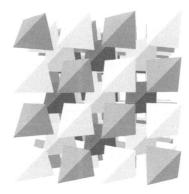

Figure 23.2. A bicubic assembly of regular tetrahedra with symmetry 4°:2.

also known as "Zometool") can only have the subgroup 4^-, obtained by restricting the point group to $3*2$, when examined in fine detail. This is because the edges, as shown at the right, have rectangular cross-section and are assembled pyritohedrally into the balls at the vertices. For the same reason, the largest of the 35 prime groups that can be exactly modeled with Zometool is $8^{-\circ}$.

Figure 23.2 shows a bicubic assembly of parallel regular tetrahedra, so (if we ignore the colors) its symmetry group (obtained by restricting the bc group to $*332$) is $4°:2$. The cyclic order is forced by the way the vertices of the tetrahedra point to each other: yellow → red → blue → green → yellow. If we pay attention to the colors, the lattice drops to fc with the same point group 332 and so the group decreases to $1°$, otherwise known as the smallest plenary group $\mathbf{p}$.

Voronoi Tilings of the Lattices

Any lattice determines a tiling of space into polyhedra called its *Voronoi cells* ("vocells"). The vocell of any lattice point consists of all points of space that are nearer to it than they are to any other lattice point. For the normal cubic (nc) lattice, the Voronoi tiling is obviously the decomposition of space into cubes (Figure 23.3, left). For the body-centered or bicubic (bc) lattice, it is the tiling ("truncoctahedrille") into truncated octahedra (Figure 23.3, right)

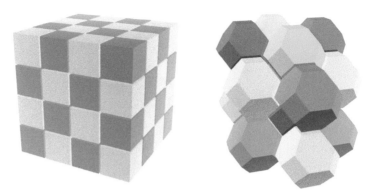

Figure 23.3. The Voronoi cells of the normal cubic lattice are cubes; those of the bicubic lattice are truncated octahedra.

described at the start of Chapter 21. The centers of the tiles are the nodes 0, 1, 2, 3 and their vertices are the nodes + and −.

Finally, for the face-centered or half-cubic (fc) lattice, the vocells are the rhombic dodecahedra of "rhombohedrille." This tiling can be obtained as follows: To obtain the vocells of the half-cubic lattice, divide alternate cubes of the lattice into six square pyramids (Figure 23.4, left) by joining their centers to their faces, and then join these square pyramids to the adjacent, undivided cubes, resulting in rhombic dodecahedra. Their centers are, say, the nodes 0, and their vertices are the other three digital nodes 1, 2, and 3 (red, blue, and green, respectively, on the right in Figure 23.4). The vertices (here 2) that have the same parity as the centers are 4-valent while those (1 and 3) of the other parity are 3-valent.

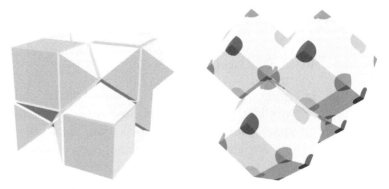

Figure 23.4. The Voronoi cells of the fc are rhombic dodecahedra.

Salt, Diamond, and Bubbles

Many simple crystals are easily described in terms of one or other of these lattices. For example, the atoms of a salt crystal will form a normal cubic lattice, but it does not have all the symmetries of the nc lattice, since the atoms alternate between sodium (Na) and chlorine (Cl)—say $1 = $ Na and $3 = $ Cl. So, we can no longer move 1 or 3, and the symmetries effect only $\langle(02),(+-)\rangle$, showing that a salt crystal has the *negative double dyad group* $2^-{:}2$.

Salt.

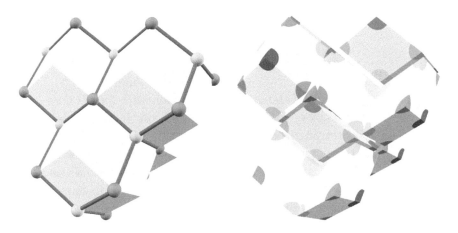

The diamond net has the same symmetry as the catoptric tiling by obcubes (Chapter 21).

A diamond crystal is made of carbon atoms situated at the 0 and 1 nodes. Its symmetries effect[1] $\langle(01)(23),(+-)\rangle$, so the group is now the *positive double dyad group* $2^+{:}2$. The diamond net is formed by the blue and green balls on the left in the figure above and the lines joining them. (Beware! The polyhedra between them are oblique cubes or "obcubes," not cubes!) The points at maximal distance from these (our red and yellow nodes on the right above, at the centers of the obcubes) would form another diamond net; the two together form what we call the "Double Diamond" with group $4^+{:}2$.

---

[1] We cannot interchange just 0 with 1 or just 2 with 3, since the permutations (01) and (23) are not in the group of possible effects.

Voronoi cells of atoms in a diamond.

The Voronoi cells of the atoms in a diamond are *triakis truncated tetrahedra* formed by erecting small pyramids on the triangular faces of a truncated tetrahedron. The Voronoi tiling is closely related to the Architectonic "trunctetrahedrille" tessellation of space into the truncated tetrahedra obtained by removing these pyramids and the ordinary tetrahedra obtained by gluing them together in sets of four. Both tessellations have the same group, 2^+:2, as the diamond.

In 1887 Lord Kelvin asked what was the most efficient way to fill space with bubbles of equal volume, and he conjectured that it was the arrangement shown on the left in Figure 23.5, in which the bubbles are "relaxed" versions of our truncated octahedra ("relaxed" because the area of the faces is reduced by allowing them to be slightly curved). We shall call these the *Scottish bubbles*.[2] Their centers are 0, 1, 2, and 3, so the group is again the double octad group $8°$:2 of symmetries of the bc lattice.

The record stood until 1993, when Weaire and Phelan found what we call the *Irish bubbles*, shown on the right in Figure 23.5, whose centers are all the points 0, 1, 2, 3, and +, by relaxing the Voronoi cells determined by these points. Their group is therefore the *pure octad group* $8°$, since + and − can no longer be interchanged.

----

[2]Lord Kelvin was actually born in Belfast: however, he did most of his work at Glasgow University and in 1892 became Baron Kelvin of Largs, the nearby town where he lived.

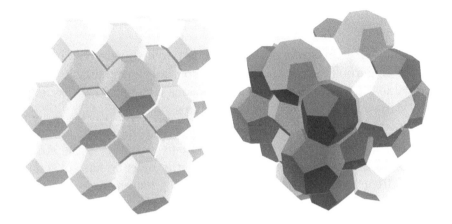

Figure 23.5. Efficient bubbles: At left, the Scottish bubbles, considered by Kelvin in 1887; at right, the more efficient Irish Bubbles, found by Phelan and Weaire in 1993. The actual bubbles are slightly "bulgy" versions of the polyhedra illustrated here.

Infinite Platonic Polyhedra

In 1923, the 16-year-old schoolboys H. S. M. Coxeter and J. F. Petrie found themselves confined to the same sanatorium and started to discuss mathematics. Later they published a paper about three new regular polyhedra [10]. These *Coxeter-Petrie polyhedra* don't seem to have received individual names in their first eight decades; since they are multiple covers of the first three Platonic solids, we shall call them the *multiplied tetrahedron, multiplied cube*, and *multiplied octahedron* and slangily shorten these to mucube (μC), muoctahedron (μO), and mutetrahedron (μT), respectively.

Any such polyhedron can be given three kinds of "orientation":

- A *1-orientation* directs lines normal to the surface.

- A *2-orientation* is an orientation in the surface.

- A *3-orientation* is an orientation of space.

Accordingly, we can speak of the 1-chiral, 2-chiral, and 3-chiral subgroups G_1, G_2, and G_3 of its symmetry group **G**. Of course, "3-chiral" is synonymous with "chiral." Fixing any two orientations fixes all three, producing a further group G_{123}, the "polychiral" part.

For the mucube (μC), we find $G_1 = 4^-{:}2$, $G_2 = 8^{\circ\circ}$, $G_3 = 8^{+\circ}$, and $G_{123} = 4^{\circ-}$.

The universal covers of these infinite polyhedra are certain tilings of the hyperbolic plane and so have hyperbolic symmetry groups, as explained in the appendix to the previous chapter. "Hyp" refers to the signatures of these hyperbolic groups.

We specify each polyhedron by its face code and a placement symbol, when appropriate. The *face code* merely lists the sizes of the faces around a vertex; for instance, the mucube has six squares around each vertex, so its face code is 4^6. *Placement symbols*, less precise, are more complicated and will be described as we go on.

The multiplied cube or mucube (μC): 4^6, Group $8^\circ{:}2$, Hyp $*\mathbf{642}$. This object—$\{4, 6 \mid 4\}$ in Coxeter's notation—has six square faces at each vertex. These are the walls of a system of (cubical) *corridors* joining empty cubical *rooms* as in the figure. The *inside* rooms are centered at nodes 0 and 2 and *outside* ones at 1 and 3. The centers of the faces are alternately $+$ and $-$. It has all the symmetries of $8^\circ{:}2$ of the bc lattice formed by the room-centers.

The multiplied octahedron (μO): 6^4, Group $8^\circ{:}2$, Hyp $*\mathbf{642}$. This—$\{6, 4 \mid 4\}$ in Coxeter's notation—has four hexagonal faces per vertex that are exactly the hexagonal faces of the truncated octahedral vocells of the bc lattice (that form the "truncoctahedrille"). Alternate vertices are the nodes $+$ and $-$. Of course, it has the same symmetries as its dual, the mucube.

The multiplied tetrahedron (μT): 6^6, Group $2^+{:}2$, Hyp $*\mathbf{662}$. The mutetrahedron—$\{6, 6 \mid 3\}$ in Coxeter's notation—has for its faces the hexagonal faces of the Voronoi tessellation for the diamond, or equivalently those of the Architectonic tessellation into truncated tetrahedra (centered at 0 and 1) and ordinary tetrahedra (centered at 2 and 3). Its group is that of the diamond, $2^+{:}2$.

It is self-dual, and its duality group in the sense of Chapters 11 is $2^+{:}2\backslash 4^+{:}2$.

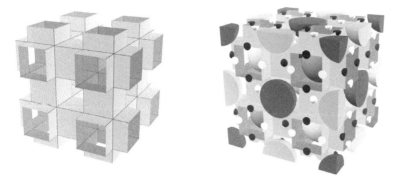

The mucube, shown with the colored nodes on the right.

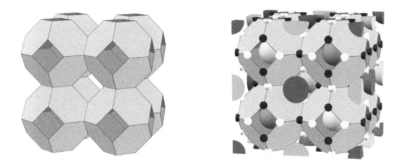

The muoctahedron.

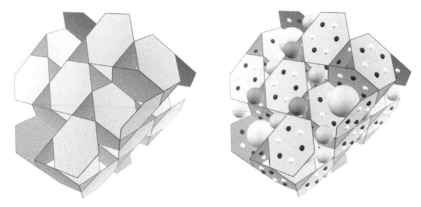

The mutetrahedron.

Their Archimedean Relatives

The next simplest class is formed by infinite polyhedra that are related to the infinite Platonic polyhedra in much the same way as the traditional Archimedean solids are related to the Platonic ones, although there are some subtleties on which we shall not elaborate.

The multiplied cuboctahedron (μCO): 6.4.6.4, Group 2^+:2, Hyp *$\mathbf{642}$. The faces are the squares of our truncated octahedra together with *half* their hexagons, namely those separating 0 and 1 or 2 and 3 but not 0 and 3 or 2 and 1.

The murhombicuboctahedron (μRCO): 6.4.4.4, Group $8°$:2, Hyp *$\mathbf{642}$. This can be obtained by exploding the faces of either μC or μO, so it has the same group.

The semi-snub mucube ($ss\mu C$): 4.3.4.3.3.3, Group $4°$:2, Hyp $\mathbf{4}$*$\mathbf{3}$. This can be obtained from the multiplied cube, regarded as $(4.4)^3$, by applying the Coxeter semi-snub operation (see Chapter 21), which replaces face code $(p.q)^r$ with $p.3.q.3.r.3$. Its square faces are the "midsquares" of those of the mucube (formed by joining the midpoints of the edges of those squares). Between them are skew hexagons that can each be filled in either of two different ways by four faces of an ordinary octahedron so as to produce a bulge on the surface. We choose these bulges so that they point alternately in and out. If the bulge between nearby nodes 0 and 1 points from 0 to 1, say $0 \rightarrow 1$, then we find this entails $1 \rightarrow 2$, $2 \rightarrow 3$, and $3 \rightarrow 0$, thus setting up the cyclic order shown below that restricts the effects to $\langle (0123), (+-) \rangle$, yielding the unusual group $4°$:2.

The propeller-hedron or double-snub mucube: 3^7, Group $4°$, Hyp $\mathbf{32}\times$. We can "oversnub" the semi-snub mucube $ss\mu C$ to obtain another fascinating polyhedron that we describe here although it really belongs with its pseudo-Platonic friends (see page 340). Divide each

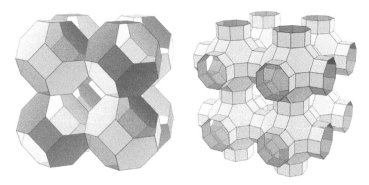

The mucuboctahedron μCO (left) and the murhombicuboctahedron μRCO (right).

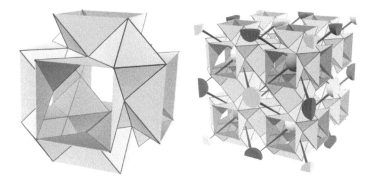

The semi-snub mucube $ss\mu C$ is shown here with colored nodes, connected to show the direction of the "bulges" described in the text. As with the bicubic lattice of tetrahedra, there is a cyclic symmetry on the colors, and the symmetry of the polyhedron is 4°:2.

$$
\begin{array}{ccc}
0 & \rightarrow & 1 \\
\uparrow & & \downarrow \\
3 & \leftarrow & 2
\end{array}
$$

Directing the axes in the semi-snub mucube.

square of $ss\mu C$ by a diagonal into two triangles in such a way that each vertex ends just one of the new diagonals. The resulting object, with nine triangles per vertex, can be deformed until all its triangles are equilateral! It turns out that sets of seven of these triangular faces are coplanar; they form propeller shapes like those of the drums in Chapter 17. The fact that the group is not generated by its finite order elements is related to the $\times$ in its hyperbolic signature.

The multiplied snub cube or musnub cube (μsC): 6.3.4.3.3, Group 8°/4, Hyp 642. This fascinating polyhedron has the face code 6.3.4.3.3 that would be produced by the traditional snub operation from 6^4 or 4^6. Its group is the largest quarter group $\mathbf{Q} = 8°/4$, and the subgroup that fixes each color in the right half of the figure is the smallest quarter group $\mathbf{q} = 1°/4$. All the other quarter groups can be obtained by restricting to the appropriate groups of color permutations.

The simplest way to understand it is to regard it as (nearly) the surface that separates black wood from white wood in checkerstix, which we define later. To keep it with its friends, we nevertheless discuss it here. The non-snub faces are the hexagons that bisect the empty cubicles between their white faces and their black ones, together with the midsquares of the squares between black and white rods. These faces are separated by skew quadrilaterals that can each be filled uniquely by two equilateral "snub" triangles. (The snub faces cut notches into the rods, which is why this only *nearly* separates the two colors.)

Smooth surfaces: Easily-described smooth surfaces approximate some of the polyhedra above. For example, the surface of points (x, y, z) that satisfy

$$\sin x + \sin y + \sin z = 0$$

closely approximates both the mucube and the muoctahedron. It even more closely approximates a famous minimal surface, the Schwarz P-surface, which can be obtained by relaxing it. This means that if the above surface were made out of soap-film, the surface tension would automatically reshape it into the Schwarz P-surface.

The centers of the star faces of the multiplied snub cube μsC lie on the bc lattice; at right, these faces are colored accordingly.

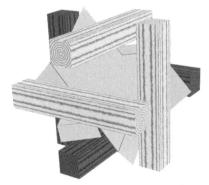

The multiplied snub cube μsC and checkerstix. The rods have been made narrow for clarity.

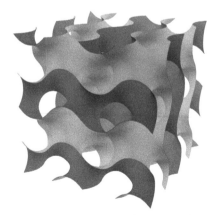

$$\sin x + \sin y + \sin z = 0 \qquad \sin x \cos y + \sin y \cos z + \sin z \cos x = 0$$

In a similar way, the points satisfying

$$\sin x \cos y + \sin y \cos z + \sin z \cos x = 0$$

closely approximate the multiplied snub cube and even more closely approximate Schoen's minimal surface, the gyroid, into which it would relax.

Pseudo-Platonic Polyhedra

We believe that nobody has yet enumerated the hundreds of "Archimedean" polyhedra in 3-space. The only further ones we'll discuss here are *pseudo-Platonic*, meaning that all their faces are the same shape.

Often they are formed by sticking together finite polyhedra, and so we name them by the principal polyhedron involved and the face code. We also give a *placement code*, which hints at how the polyhedra are placed and joined, and also makes it easy to find the group in a way we shall describe with our first example.

It turns out that most of these polyhedra are 1-chiral, because their "insides" and "outsides" are different shapes. So, their groups have only one "chiral part," the chiral subgroup G_3. The propeller-hedron that starts this chapter is the exception.

Icosahedral 3^7: Group $2^{\circ+}$, Hyp **3222**, Chiral part 1°, Placement $I\{0_+, 1_-\}$ with O-joins. The placement code here means that we place icosahedra in the $+$ orientation at each node 0, and in the $-$ orientation at each 1, choosing their sizes so that adjacent ones can be joined by an octahedron. The result is that four octahedra join each 0-icosahedron to four parallel 1-icosahedra. The group is found by asking what permissible permutations of 0, 1, 2, 3, $+$, and $-$ fix the code. Since 0 and 1 can only be interchanged if 2 and 3 are, there is only one nontrivial one, namely $(01)(23)(+-)$, and so the group is $2^{\circ+}$.

Octahedral 3^8: Group $2^+{:}2$, Hyp **2∗32**, Chiral part 2^+, Placement $O\{0, 1\}$ with O-joins. The same, with octahedra replacing icosahedra. Since these have only one orientation, we can now separately perform $(01)(23)$ and $(+-)$, so the group is $2^+{:}2$.

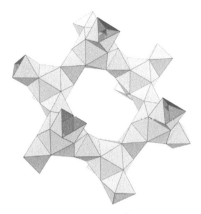

Icosahedral 3^7.

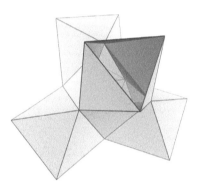

Octahedral 3^8.

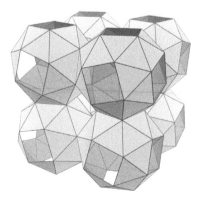

Snub-cubical 3^8.

Snub-cubical 3^8: Group 4^{--}, Hyp **32∗**, Chiral part $2^{\circ-}$, Placement $sC\{0_L, 2_R\}$, touching. The placement code means that we inscribe snub cubes in the cubes of the normal cubic tiling, of alternate handedness (left-handed L around 0-nodes, right-handed R around 2-nodes). Since odd permutations interchange L and R, we can understand the group by writing its four effects as

$$\text{identity}, (13)(+-), (02)(LR), (02)(13)(+-)(LR).$$

So (dropping this convention), it is $\langle(13)(+-), (02)\rangle = 4^{--}$.

This is the unique group that is semisplit in two distinct ways. The local groups at the centers 0 and 2 are the chiral cubic group **432** of the snub cube corresponding to our placements 0_L and 2_R. However, the structures at the vertices are pyritohedral **3∗2**, and the alternative placement code $\{1_+, 3_-\}$ describing these gives an easier way to compute the group as the 4^{--} generated by (02) and $(13)(+-)$.

Icosahedral 3^9: Group $8^{-\circ}$, Hyp **22∗2**, Chiral part $4^{\circ\circ}$, Placement $I\{0_+, 1_-, 2_+, 3_-\}$, with O-joins. Now each icosahedron is joined to eight others by octahedral rings. Plainly the group 4^- generated by (02) and (13) fixes the code, as does $(01)(23)(+-)$, which extends this to the hybrid group $8^{-\circ}$.

Octahedral 3^{12}: Group $8^{\circ}{:}2$, Hyp **2∗42**, Chiral part $8^{+\circ}$, Placement $O\{0, 1, 2, 3\}$, with O-joins. The same with octahedra for icosahedra. Plainly its group contains all permissible permutations of $0, 1, 2, 3$ and $+, -$.

Cubical 4^5: Group $8^{\circ}{:}2$, Hyp ∗**4222**, Chiral part $8^{\circ+}$, Placement $C\{0, 1, 2, 3\}$ with $\frac{1}{27}$ overlap. The "cubes" of the placement code are empty $3{\times}3{\times}3$ cubical chambers, of which adjacent ones intersect in $1 \times 1 \times 1$ cubes.

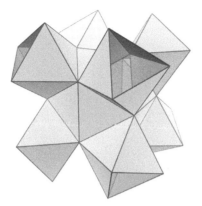

Icosahedral 3^9.

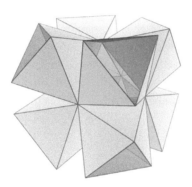

Octahedral 3^{12}.

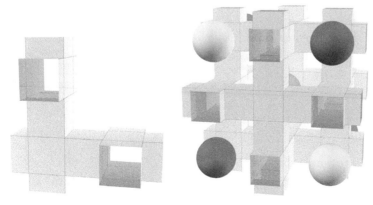

Cubical 4^5.

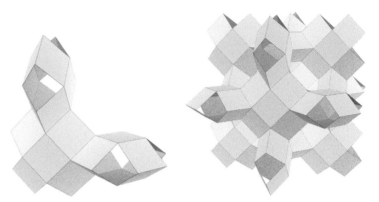

Truncoctahedral 4^5.

Truncoctahedral 4^5: Group $8°{:}2$, Hyp **2∗42**, Chiral part $8°^+$, Placement tO{0,1,2,3} with $P6$-joins. The placement code indicates that this can be obtained by joining truncated octahedra by hexagonal prisms. The truncated octahedra are the large chambers in the figure, but they are not so easy to see because their hexagons are missing. By comparing the small piece shown with the corresponding piece of the cubical 4^5, we see that the two 4^5 polyhedra have the same symmetries.

The Three Atomic Nets and Their Septa

There are, in fact, three interesting nets made by line segments joining suitable points, which we'll call *atoms*:

- the *unit net*, formed by the edges of the tiling by unit cubes, whose atoms are those of a salt crystal;

- the *dia-net*, formed by the atoms and valence bonds of a diamond;

- the *tria-net*, which we shall define later in this chapter.

These are analogous in many ways. For example, the points maximally distant from the atoms of one of these nets are the atoms of another net of the same type—its mate, which together with the original form a "double-net" with twice the symmetry. For example, the $4^- : 2$ of the unit net becomes $8° : 2$ for the double unit net; similarly, $2^+ : 2$ becomes $4^+ : 2$ for the double diamond, and $4^+/4$ becomes $8°/4$ for the double triamond.

For each of the three double nets, there is a natural family of surfaces, all topologically equivalent, that separate them. One such surface is formed by the points maximally distant from the double net, and the others are topological deformations of this. We call any such surface a *unit-septum*, *dia-septum*, or *tria-septum*, respectively (since the word "septum" means "separating surface").

Most of the infinite polyhedra that we've discussed are septa of these three types. Thus, the mucube and the muoctahedron are unit-septa, while the mucuboctahedron is dia-septal and the musnub cube tria-septal. This is no accident: it can be shown that any surface that supports[3] a hyperbolic triangle group (i.e., one of shape **pqr**, ∗**pqr**, or **p**∗**q**) is a septum of one of these three types.

Naming Points

We have already used notations like **12** for the midpoint of the segment joining a **1**-node to a **2**-node. We can extend this to notations like $\mathbf{1}^2\mathbf{2}$ for the point one third of the way along such a segment. In general if **0**, **1**, **2**, and **3** are the vectors to the nodes of a tetrahedron of sign ±, then we write $\mathbf{0}^a\mathbf{1}^b\mathbf{2}^c\mathbf{3}^d$ (+ or −) for the point of the tetrahedron represented by $(a\mathbf{0} + b\mathbf{1} + c\mathbf{2} + d\mathbf{3})/(a+b+c+d)$. There are obvious abbreviations when one of a, b, c, d is 0 or 1, and we can omit the sign when it doesn't matter (which it certainly won't if one of a, b, c, d is 0, since then the point lies on the surface of tetrahedra of both kinds).

Often we're not interested in the exact values of a, b, c, d but just their relative sizes. In this case, we can abbreviate $\mathbf{0}^a\mathbf{1}^b$, $a > b$, to $\mathbf{0_1}$ and $\mathbf{0}^a\mathbf{1}^b\mathbf{2}^c$, $a = b > c$, to $\mathbf{01_2}$, for example.

Any set of equivalent points under one of the plenary groups can be named in this way. For the quarter groups, we add a further convention. The directions of the triad axes we name according to

---

[3]Strictly, the lifts of its symmetries to the universal cover form such a group.

their odd coordinate out as

$$w = (+++), x = (-++), y = (+-+), z = (++-),$$

and we append these letters as subscripts to the names of the nodes. For instance, $\mathbf{2}_z$ denotes a node $\mathbf{2}$ that lies on a triad axis in direction $(++-)$.

The nodes of the triamond are, thus, $\mathbf{0}_v\mathbf{1}_v$, $\mathbf{1}_v\mathbf{2}_v$, $\mathbf{2}_v\mathbf{3}_v$, and $\mathbf{3}_v\mathbf{0}_v$, where v denotes any one of w, x, y, z.

We usually understand such names only up to even permutations of w, x, y, z, since these lead to points that are equivalent under all the quarter groups. Thus, squares of the musnub cube can be described simply as having centers at $\mathbf{0}_w\mathbf{1}_x\mathbf{2}_y\mathbf{3}_z$ and vertices $\mathbf{0}_w\mathbf{1}_x^2\mathbf{2}_y$, $\mathbf{1}_x\mathbf{2}_y^2\mathbf{3}_z$, $\mathbf{2}_y\mathbf{3}_z^2\mathbf{0}_w$, and $\mathbf{3}_z\mathbf{0}_w^2\mathbf{1}_x$.

Polystix

There's a nice way to arrange square-section sticks (*tetrastix*) so as to occupy 75% of space. The unoccupied space consists of $\frac{1}{2} \times \frac{1}{2} \times \frac{1}{2}$ cubes centered at nodes 0, 1, 2, and 3 and surrounded "pyritohedrally" as shown below. Remarkably, there is an alternative system of tetrastix that occupies exactly the same portion of space. The two systems may be distinguished by the directions of the grain of the wood abutting the faces of the empty cubes, so we'll call them +-tetrastix and −-tetrastix. The effects of the symmetries of either are just the permutations that fix + and −, so their symmetries are the pure octad group 8°.

A precisely similar thing happens for sticks whose cross-sections are equilateral triangles (*tristix*). This time the holes are rhombic dodecahedra of the four colors, and again there are two arrangements of tristix that occupy the same space.

We indicate them by $\{0_L1_L2_L3_L\}$ or $\{0_R1_R2_R3_R\}$, since the holes are at 0, 1, 2, and 3 and the grains around a hole are either all left-handed or all right-handed. Their common group is the chiral portion of 8°:2, namely $8^{-\circ}$.

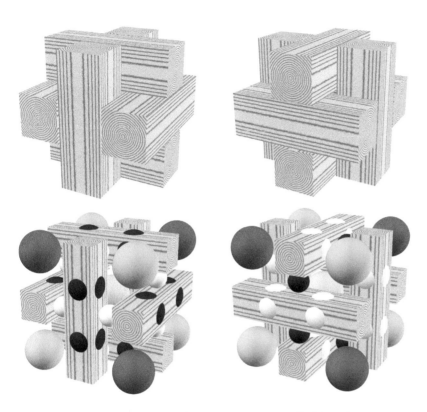

How two systems of tetrastix occupy the same space (group 8°). Below, the empty spaces contain nodes of four different colors, of which only three can be seen here.

Tristix (group $8^{-\circ}$): At left, a bc lattice of rhombic dodecahedral "holes"; there are two ways to fill in tristix, $\{0_L 1_L 2_L 3_L\}$ (middle) and $\{0_R 1_R 2_R 3_R\}$ (right).

The hexastix arrangement (of which a model made with pencils is shown below) is more subtle, and we shall postpone its discussion for a moment).

Hexastix: If these pencils (after an idea of George Hart) were infinitely long we'd get hexastix (group $8°/4$); if we sharpened them, we'd get hexastakes, which has group $4^+/4$.

Checkerstix and the Quarter Groups

How many ways can we color the sticks of a tetrastix model so that those of any given direction are alternately black and white, like a checkerboard? Since there are two choices for each of three directions, the answer is $2^3 = 8$, but we reduce this to 4 by regarding the overall interchange of black and white as giving "the same" checkering.

All these checkerings are related by elements of the tetrastix group $8°$, so the group that fixes any one of them contains just one quarter of the elements of that group. We therefore call it the *quarter octad group* $8°/4$. By intersecting it with the remaining pure groups 4^-, $4°$, 4^+, 2^-, $2°$, 2^+, and $1°$, we obtain the other quarter groups, respectively $4^-/4$, $4°/4$, $4^+/4$, $2^-/4$, $2°/4$, $2^+/4$, and $1°/4$.

Checkerstix and hexastix: group $8°/4$, or $4^+/4$ if we fix black and white.

Hexastix from Checkerstix

It is time to define hexastix more precisely!

Each empty cubicle of checkerstix has a natural *diagonal*, from the vertex on three white faces to the opposite one, on three black ones (shown by white rods in the figure above). These diagonals join up to produce the triad axes of the quarter octad group $8°/4$, the empty cubicles being skewered along them like meat on a kebab, and the convex hulls of these kebabs are the hexagonal "pencils" of hexastix.

Tristakes, Hexastakes, and Tetrastakes

We get some subgroups by assigning a direction to each stick of tetrastix, tristix, or hexastix. In figures this can be indicated by sharpening one end of the stick, turning it into a "stake."

This naturally gives rise to two groups, a smaller one that takes all sharpened ends to sharpened ends, and a larger one that also includes operations that interchange sharpened ends with unsharpened ones. If the larger one is called $\mathbf{G}$, then the smaller one is its 1-chiral part $\mathbf{G}_1$. The ordinary chiral part is again called $\mathbf{G}_3$, and there is a 2-chiral part preserving the orientation of the planes perpendicular to the sticks and a polychiral part $\mathbf{G}_{123}$ that preserves all three orientations.

Tristakes (group 4°°).

For tristakes and hexastakes, there's only one sensible way to choose the directions, and it's equivalent to restricting the point group to **332** and ∗**332**, respectively, which gives the groups 4°° and $4^+/4$, respectively.

For tetrastakes, we have a remarkable coincidence: "staking" a tetrastix, up to reversing all directions, is equivalent to "checkering" its mate, up to interchanging all colors. To understand this, look at the figure below. It shows how the (wooden) rods of one tetrastix cut those (the white "pencils") of its mate. If these pencils have been sharpened into "stakes" (directed alternately), then each wooden rod acquires a spiral structure. In the figure we have made it of pine if the spiral is *dextral* (like a corkscrew) and of ebony if it is *sinistral* (a reflected corkscrew).

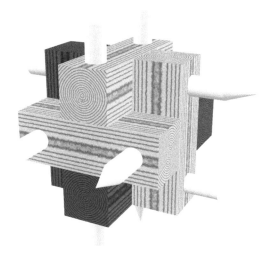

Understanding the Irish Bubbles

The Irish bubbles are closely related to tetrastix: regard each rod of tetrastix as divided into cubes in the usual way, and then push all these divided rods one half of a cube-width along themselves. Then, if the cube-surfaces were relaxed, they would become the two shapes of Irish bubble: a dodecahedron (with only pyritohedral symmetry) and a *dodecadihedron*, so called because it has 12 ("dodeca") pentagonal faces and 2 ("di") hexagonal ones. So, tetrastix and the Irish bubbles have the same symmetry (8°)—the original empty holes are now dodecahedral, while the dodecadihedral bubbles stack along their hexagons into nubbly columns that are what the rods have become.

The Triamond Net and Hemistix

The net $[10, 3]_a$ was given its name by A. F. Wells in the 1970s. Its existence is implicit in a paper written by Laves in the first half of the twentieth century, and it was discussed explicitly by Coxeter in the 1950s. It was rediscovered as recently as 2007 by Sunada; it is discussed by Ludwig Danzer in [13].

It is closely related to the object we call *hemistix*, formed by the rods of one color in checkerstix. In hemistix, the rods that touch a given one spiral around it (the spirals being dextral for one color, sinistral for the other.) Below, we show a segment from each such rod, and the positions of the nodes 0, 1, 2, and 3.

Hemistix and the Triamond: On the left, hemistix, one half of the checkerstix. In the middle, we see how neighboring rods coil around a given rod. At right, a portion of the Triamond.

The segments of the Triamond wind helically around each original stick, in the squares where it touches its neighbors. Thus, a Triamond, like its parent hemistix, is either dextral or sinistral and has for its group the chiral part of $8°/4$, namely the *positive quarter tetrad group* $4^+/4$. The points of space at maximal distance from one Triamond form another, of the opposite handedness, coming from the other half of checkerstix: the two Triamonds together form a "Double Triamond" (compare Double Diamond), with group $8°/4$. We close this section with a three-dimensional view of the Triamond for those lucky enough to be able to view stereoscopic images, with or without a stereoscope.

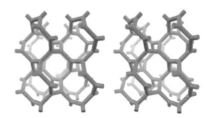

The Triamond net (group $4^+/4$).

Further Remarks about Space Groups

The three-dimensional analogs of the 17 plane crystallographic groups are the 219 (three-dimensional) space groups. This chapter and the last have been about the 35 most interesting ones, namely the "prime" ones that don't fix any family of parallel lines; the "composite" ones that do fix such a family are handled in Chapter 25.

We have chosen to distinguish the two classes by these terms because the "composite" space groups are those that are in a certain sense composed of lower-dimensional ones. Crystallographers usually use the term "cubic" for our "prime" because the point groups of the 35 prime groups are all contained in $*\mathbf{432}$, the symmetry group of the cube. However, since other groups have this property, this term is insufficient. The true criterion, already sufficient in itself, is that these groups do not fix any direction, i.e., are prime in our sense.

Crystallographers usually give the number of space groups as 230. This is because 11 of them have both left-handed and right-handed forms, that is to say, they are *metachiral*. Metachirality is stronger than mere chirality. We say an object is *chiral* if it differs from its mirror image, and traditionally we also say that its symmetries form a chiral group. However, the group itself might still be the same as *its* mirror image! When it is *not*, the group is metachiral.

In the real world that chemists inhabit, we cannot interchange left and right, so for their purposes the count of 230 is quite correct. However, as mathematicians, we prefer to obey Thurston's commandment that one should always think of groups in terms of their orbifolds, and this makes 219 the appropriate number.

What happens is that the orbifolds of the 11 metachiral groups each have two orientations. However, many orbifolds have no orientation, with the consequence that there is no way of counting orbifolds that yields the number 230.

– 24 –

Flat Universes

What is the shape of the universe? For a long time, people thought that it had to be the infinite three-dimensional space described by Euclid and therefore called it *Euclidean 3-space*. Then a few people realized that that wasn't the only possibility. It might, for instance, be like the surface of a sphere in four-dimensional space, which nowadays is called three-dimensional *spherical space* or just the *3-sphere*. Then, it would be "finite but unbounded," meaning that the total volume would be finite even though there would be no bounding walls. In that case, space would be curved, in a three-dimensional analog to the two-dimensional surface of the Earth, which has finite total area even though there is no edge for people to fall off. Still more recently, physicists have become aware that the universe might not only be finite and unbounded, but also exactly *flat*, in the sense that the local geometry is exactly Euclidean.

It might, for instance, be the three-dimensional analog of a torus, which we shall call a *torocosm*. If you've played computer games such as *Asteroids*, you won't find it too hard to understand. In that game, an asteroid or space ship that goes off one edge of the screen reappears from the opposite edge.

Flying around in an *Asteroids* universe: As the ship exits one side of the picture of its universe, it reappears on the other side. From the ship's perspective, the universe is unbroken and continuous. This universe is finite, yet unbounded, and is topologically a torus.

(opposite page) The Hexacosm.

This chapter is about all the ways the universe might be flat. First we'll think about the two-dimensional analogs.

Compact Platycosms

If you lived in a small, flat two-dimensional space like the *Asteroids* universe, what would you see? Because light rays travel like everything else in that world, you'd see lots of images of any given thing.

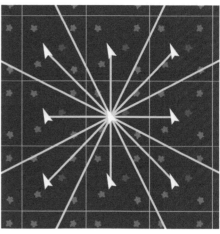

Sight in the *Asteroids* universe: Light would travel like any other thing, wrapping around the universe. From the point of view of the ship, it would seem as if the light travelled across consecutive *copies* of the universe, and you would see lots of images of any given thing…including yourself! These copies form the *universal cover* of the torus, which is a Euclidean plane.

The entire set of images you'd see would be related to each other by a lattice of translations. Geometrically, these translations form a copy of the plane group we call ○, and the flat torus is just the orbifold for ○.

Thus, flat universes are described by space groups, in that they describe ways in which one might return to one's starting place after "circumnavigating the universe." We shall obtain the ten most interesting flat universes from ten of the space groups described in the next chapter. Technically these are the ten closed flat 3-manifolds.

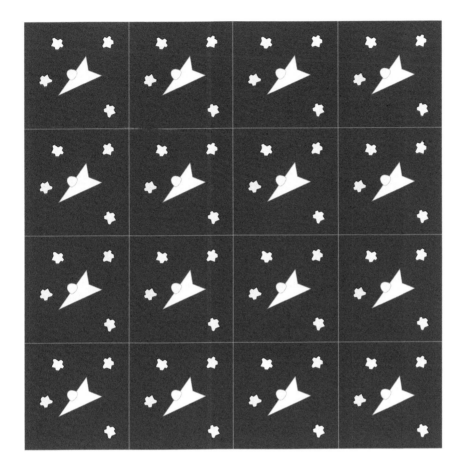

Torocosms

The above figure represents the appearance of a two-dimensional torus. A (flat) *torocosm* (also called a "3-torus") is its three-dimensional analog. It's like a room without walls with the strange property that if you leave from one side you come in at the opposite side.

The figure on the next page suggests what the universe would look like if it were a sufficiently small torocosm. Geometrically, a torocosm is the orbifold of the space group generated by three independent translations. These need not be mutually perpendicular—there is a torocosm based on any parallelepiped.

The Klein Bottle as a Universe

Games like *Asteroids* can also be played on other surfaces, for example, on a flat Klein bottle.

Flying around in a Klein-bottle universe: Now, whatever leaves the screen on the top will reenter it on the bottom, but reverse. However, what leaves at the right returns in the way it did in the original game.

Geometrically, this twisted Asteroids world is of course a Klein bottle, the orbifold of the plane group ××. The torus and Klein bottle are the only closed, flat 2-manifolds. Why is this? Because the universal cover argument shows that any such space must be the

orbifold of one of the 17 plane groups, and each of the 15 other than
○ and ×× has either a boundary (if its name involves ∗) or a cone
point, at which the local geometry will not be Euclidean.

The Other Platycosms

The three-dimensional analogs of the torus and Klein bottle we call
platycosms, meaning "flat universes." More precisely, they are the
compact platycosms, meaning that they have finite volume. There
are ten such spaces [8], whose individual names are

> torocosm, dicosm, tricosm, tetracosm, hexacosm

(these are collectively called the *helicosms*) and

> didicosm, positive and negative amphicosms,
> and positive and negative amphidicosms.

We describe them by pictures. We have already seen the hexacosm
(on page 354) and the torocosm (previous page), which is the helicosm with $N = 1$. In another helicosm, say the N-cosm, the images
of an object fall into parallel layers and one moves up a layer by
rotating it through $360°/N$.

The images of any one layer are related to each other by a lattice
of translations that has this rotation as a symmetry—for instance,
it must be a square lattice for the tetracosm.

So, in an N-cosm, one's images face in just N different directions,
which are related by a cyclic group of N rotations. They all have
the same handedness: if *you* were to wave your right hand, each of
your images would wave *its* right hand, in contrast with what would
happen in a mirror, where your image would wave its left hand. This
is described technically by saying that the helicosms are all *orientable*
manifolds.

The only other orientable platycosm is the *didicosm*, in which
one's images are in four different orientations. They can be head
up or feet up and face either left or right, but are all of the same
handedness.

The dicosm.

The tricosm.

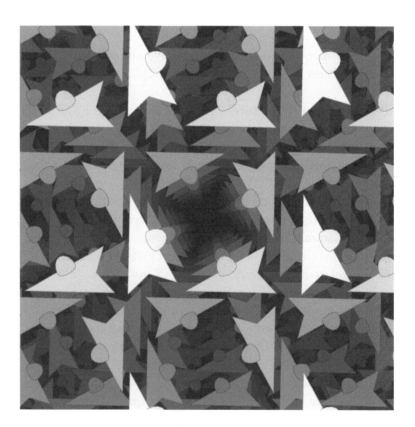

The tetracosm.

We shall show the remaining platycosms with ships as before, but for clarity we will also use blocks that have the letters b, d, p, or q on them; each kind of block stacks top and bottom, right and left, to form layers, but the letters are always on the tops of the blocks.

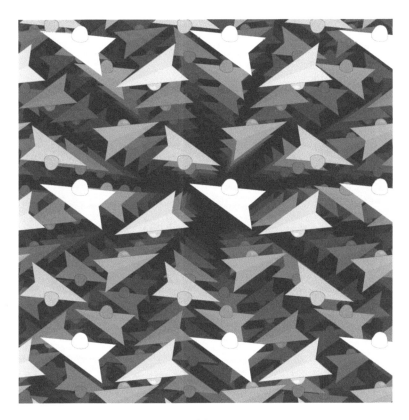

The didicosm.

The *amphicosms* and *amphidicosms* are nonorientable, meaning that half the images of an object have one handedness and half the other. So, if you were to wave your right hand in one of these four platycosms, half of your images would wave their right hands, the other half their left hands. It is hard to illustrate these correctly (the four pictures in [31] represent only two manifolds). The cockpits of the ships in our pictures are always on top.

The amphicosms and amphidicosms: from left to right, +a1, −a1, +a2, and −a2.

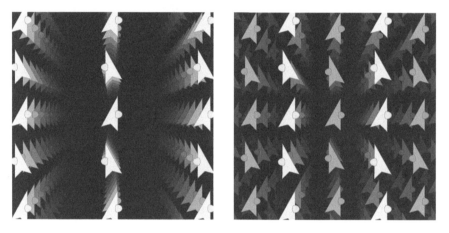

The positive amphicosm +a1.

The negative amphicosm -a1.

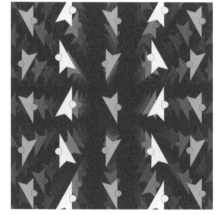

The positive amphidicosm +a2.

The negative amphidicosm -a2.

The reason that these ten platycosms are the only compact ones is much the same as the reason that the torus and Klein bottle are the only compact two-dimensional flat surfaces, namely that the corresponding space groups are the only ones in which there is no fixed point for any nonidentical element. This is easily checked by hand from the list in the next chapter. The 35 prime space groups have lines that are analogous to order-3 cone points in a plane group and so need not be considered.

The ten space groups that arise are those called

$$
\begin{array}{ll}
\text{c1} & (\circ), \\
\text{c2} & (2_1 2_1 2_1 2_1) = (\bar{\times}\,\bar{\times}), \\
\text{c3} & (3_1 3_1 3_1), \\
\text{c4} & (4_1 4_1 2_1), \\
\text{c6} & (6_1 3_1 2_1), \\
\text{c22} & (2_1 2_1 \bar{\times}), \\
+\text{a1} & (\bar{\circ}_0) = (*{:}*{:}) = (\times\times_0), \\
-\text{a1} & (\bar{\circ}_1) = (*{:}\times) = (\times\times_1), \\
+\text{a2} & (2_1 2_1 *{:}) = (\bar{*}{:}\bar{*}{:}) = (\bar{\times}\times_0), \\
-\text{a2} & (2_1 2_1 \times{:}) = (*{:}\bar{\times}) = (\bar{\times}\times_1),
\end{array}
$$

in the next chapter. Some of these groups have more than one name as a consequence of the alias problem discussed there. We summarize the names and notations for the ten compact platycosms below.

	helicosms	didicosm
orientable	c1 c2 c3 c4 c6	c22
nonorientable	+a1 -a1 amphicosms	+a2 -a2 amphidicosms

Infinite Platycosms

The compact platycosms are the only ones of finite volume, but there are eight more infinite, or noncompact, ones. The infinite two-dimensional flat surfaces are the Euclidean plane, the cylinder, and the Möbius cylinder.

The following figures suggest what it would look like to live in these spaces. The infinite platycosms are named in the captions.

Circular Prospace $\cong$ circle x Euclidean plane.

Circular Mospace.

Toroidal Prospace $\cong$ torus x Euclidean line.

Toroidal Mospace.

Kleinian Prospace $\cong$ Klein bottle x Euclidean line.

Orientable Kleinian Mospace.

Nonorientable Kleinian Mospace.

Euclidean three-space.

Here, "Prospace" abbreviates "product space" because those three spaces are the direct products indicated. "Mospace" means "Möbius space": the four Mospaces are related to the Prospaces in the way that the Möbius cylinder relates to the ordinary one.

Where Are We?

This chapter has described the three-dimensional flat universes (platycosms). The enumeration of the ten compact ones was deduced from that of the 184 composite space groups in the next chapter.

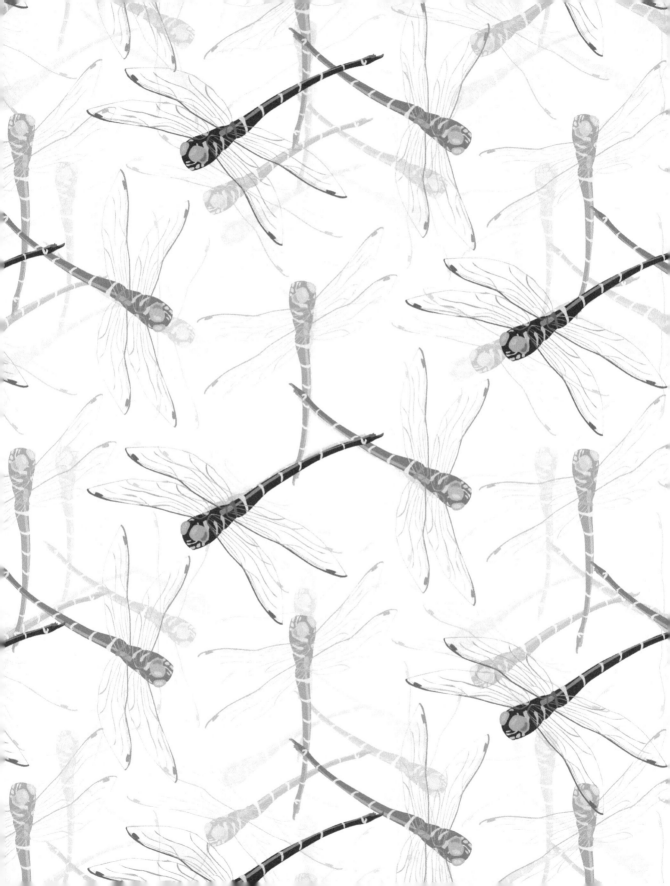

– 25 –

The 184 Composite Space Groups

The *composite* space groups are the groups that preserve at least one family of parallel lines, and so, in a suitable sense, are composed of one- and two-dimensional groups, which makes them less interesting than the 35 prime groups of Chapter 22. For that reason, we shall only explain how to read Tables 25.1–25.17, taken from [6].

Let's suppose that the invariant direction is the vertical or z-direction. Then, if we look downwards from a great height while some group operation is being performed, we will see an operation of one of the 17 plane crystallographic groups, which we shall call the *horizontal group* for this case. Each operation of this group is coupled to a *vertical* operation of the form $z \to k+z$ or $z \to k-z$.

We suppose that the smallest upward translation in the group is $z \to 1+z$, so that the vertical translations are $z \to n+z$ for all integers n. Modulo these, any other vertical operation takes the form $z \to f+z$ or $f-z$, where $0 \leq f < 1$. We'll abbreviate these to $f+$ and $f-$.

Tables 25.1–25.17 merely specify, in this notation, the possible ways that the generators of a given one of the 17 groups can be coupled with such vertical operations.

Thus, for instance, the entry $\frac{1}{2}+\frac{1}{3}-0-$ for the group whose "fibrifold name" is $(* : 63_12)$ tells us that this group is obtained from the horizontal group $*632 = *^P6^Q3^R2$, by coupling the generators P, Q, and R with $z \to \frac{1}{2}+z$, $z \to \frac{1}{3}-z$, and $z \to -z$, respectively.

The *fibrifold name* is part of a "decorated form" of the name of the horizontal group, according to the system described in [6],

---

(opposite page) The space group denoted $(3_13_13_1)$, with coupling $\frac{1}{3}+\frac{1}{3}+\frac{1}{3}+$, is the tricosm.

367

Figure 25.1. The space group $(*{:}6{:}3{:}2)$, with coupling $\frac{1}{2}+\frac{1}{2}+\frac{1}{2}+$.

inside a kind of bracket that specifies what is coupled to the identity. The parentheses used in this case tell us that these are merely the translations $n+$. Square brackets would indicate both these and the reflections, $n-$, for all integers n.

For example, the entries $\frac{1}{2}+\frac{1}{2}+\frac{1}{2}+$ and $\frac{1}{2}+\frac{1}{2}+\frac{1}{2}+0-$ for $(*{:}6{:}3{:}2)$ and $[*{:}6{:}3{:}2]$ (Figures 25.1 and 25.2) show that P, Q, and R are all coupled to $z \to \frac{1}{2}+z$ in both these groups, but the latter also contains $z \to -z$, since the identity is coupled to $0-$.

The Alias Problem

Some groups have more than one fibrifold name, because they have more than one invariant direction. In fact there are never more than three names, and all the names for a given group are collected in the following tables, in which the 184 composite space groups are organized by point group.

Figure 25.2. The space group $[*:6:3:2]$, with coupling $\frac{1}{2}+\frac{1}{2}+\frac{1}{2}+0-$.

This is all you should need to identify any of the groups listed, but some readers will want to understand roughly how the lists were proved to be complete. The idea is to find all ways of "decorating" the symbols in the orbifold notations so as to indicate the topological features of the "fibered orbifold," or *fibrifold*. This is a delicate business, because it is important that whatever is written should be an invariant feature of the fibrifold.

For example, we use a subscripted digit n_d when a horizontal rotation of order n couples to $\frac{d}{n}+$, but merely n when it couples to any $k-$, since the value of k is not an invariant. Again, when the space between two digits corresponds to a reflection, we fill it with · or : when this relfection is coupled to $0+$ or $\frac{1}{2}+$, but leave it blank for a coupling to any $k-$, for a similar reason.

We refer the reader to the original paper [6] for the coupling rules for the further generators corresponding to the symbols $*$, ∘, and ×, since they are rather subtle and not needed to understand the tables.

Fibrifold name	Couplings for $P\ Q\ R\ (I)$	Point group	Intern. no.
[*·6·3·2]	$0+0+0+0-$	*226	191
[*:6·3·2]	$\frac{1}{2}+0+0+0-$	*226	194
[*·6:3·2]	$0+\frac{1}{2}+\frac{1}{2}+0-$	*226	193
[*:6:3·2]	$\frac{1}{2}+\frac{1}{2}+\frac{1}{2}+0-$	*226	192
(*·6·3·2)	$0+0+0+$	*66	183
(*:6·3·2)	$\frac{1}{2}+0+0+$	*66	186
(*·6:3·2)	$0+\frac{1}{2}+\frac{1}{2}+$	*66	185
(*:6:3·2)	$\frac{1}{2}+\frac{1}{2}+\frac{1}{2}+$	*66	184
(*·6·3·2)	$0-0+0+$	2*3	164
(*·6:3·2)	$0-\frac{1}{2}+\frac{1}{2}+$	2*3	165
(*·6 3_0 2)	$0+0-0-$	2*3	162
(*·6 3_1 2)	$0+\frac{1}{3}-0-$	2*3	166
(*:6 3_0 2)	$\frac{1}{2}+0-0-$	2*3	163
(*:6 3_1 2)	$\frac{1}{2}+\frac{1}{3}-0-$	2*3	167
(*$6_0 3_0 2_0$)	$0-0-0-$	226	177
(*$6_1 3_1 2_1$)	$\frac{1}{2}-\frac{1}{3}-0-$	226	178
(*$6_2 3_2 2_0$)	$0-\frac{2}{3}-0-$	226	180
(*$6_3 3_0 2_1$)	$\frac{1}{2}-0-0-$	226	182

Table 25.1. Plane group: *632; Relations $*^{P}6\,^{Q}3\,^{R}2$: $1 = P^2 = (PQ)^6 = Q^2 = (QR)^3 = R^2 = (RP)^2$

Fibrifold name	Couplings for $\gamma\ \delta\ \epsilon\ (I)$	Point group	Intern. no.
[$6_0 3_0 2_0$]	$0+0+0+0-$	6*	175
[$6_3 3_0 2_1$]	$\frac{1}{2}+0+\frac{1}{2}+0-$	6*	176
($6_0 3_0 2_0$)	$0+0+0+$	66	168
($6_1 3_1 2_1$)	$\frac{1}{6}+\frac{1}{3}+\frac{1}{2}+$	66	169
($6_2 3_2 2_0$)	$\frac{1}{3}+\frac{2}{3}+0+$	66	171
($6_3 3_0 2_1$)	$\frac{1}{2}+0+\frac{1}{2}+$	66	173
($6\ 3_0\ 2$)	$0-0+0-$	3×	147
($6\ 3_1\ 2$)	$\frac{1}{3}-\frac{1}{3}+0-$	3×	148

Table 25.2. Plane group: 632; Relations $\gamma^6\,{}^{\delta}3\,{}^{\epsilon}2$: $1 = \gamma^6 = \delta^3 = \epsilon^2 = \gamma\delta\epsilon$

Fibrifold name	Couplings for $P\ Q\ R\ (I)$	Point group	Intern. no.
[*·4·4·2]	$0+0+0+0-$	*224	123
[*·4·4·2]	$0+0+\frac{1}{2}+0-$	*224	139
[*·4·4·2]	$0+\frac{1}{2}+0+0-$	*224	131
[*·4·4·2]	$0+\frac{1}{2}+\frac{1}{2}+0-$	*224	140
[*·4·4·2]	$\frac{1}{2}+0+\frac{1}{2}+0-$	*224	132
[*·4·4·2]	$\frac{1}{2}+\frac{1}{2}+\frac{1}{2}+0-$	*224	124
(*·4·4·2)	$0+0+0+$	*44	99
(*·4·4·2)	$0+0+\frac{1}{2}+$	*44	107
(*·4·4·2)	$0+\frac{1}{2}+0+$	*44	105
(*·4·4·2)	$0+\frac{1}{2}+\frac{1}{2}+$	*44	108
(*:4·4·2)	$\frac{1}{2}+0+\frac{1}{2}+$	*44	101
(*:4:4·2)	$\frac{1}{2}+\frac{1}{2}+\frac{1}{2}+$	*44	103
(*·4·4·2)	$0-0+0+$	*224	129
(*·4·4·2)	$0-0+\frac{1}{2}+$	*224	137
(*·4·4·2)	$0-\frac{1}{2}+0+$	*224	138
(*·4·4·2)	$0-\frac{1}{2}+\frac{1}{2}+$	*224	130
(*·4 4·2)	$0+0-0+$	2*2	115
(*·4 4·2)	$0+0-\frac{1}{2}+$	2*2	121
(*:4 4·2)	$\frac{1}{2}+0-\frac{1}{2}+$	2*2	116
(*$4_0$4·2)	$0-0-0+$	*224	125
(*$4_1$4·2)	$\frac{1}{4}-0-0+$	*224	141
(*$4_2$4·2)	$\frac{1}{2}-0-0+$	*224	134
(*$4_0$4·2)	$0-0-\frac{1}{2}+$	*224	126
(*$4_1$4·2)	$\frac{1}{4}-0-\frac{1}{2}+$	*224	142
(*$4_2$4·2)	$\frac{1}{2}-0-\frac{1}{2}+$	*224	133
(*·4·4 2_0)	$0-0+0-$	2*2	111
(*·4·4 2_1)	$\frac{1}{2}-0+0-$	2*2	119
(*:4:4 2_0)	$0-\frac{1}{2}+0-$	2*2	112
(*:4:4 2_1)	$\frac{1}{2}-\frac{1}{2}+0-$	2*2	120
(*$4_0 4_0 2_0$)	$0-0-0-$	224	89
(*$4_1 4_1 2_1$)	$\frac{1}{2}-\frac{1}{4}-0-$	224	91
(*$4_2 4_2 2_0$)	$0-\frac{1}{2}-0-$	224	93
(*$4_2 4_0 2_1$)	$\frac{1}{2}-0-0-$	224	97
(*$4_3 4_1 2_0$)	$0-\frac{1}{4}-0-$	224	98

Table 25.3. Plane group: *442; Relations $*^{P}4\,^{Q}4\,^{R}2$: $1 = P^2 = (PQ)^4 = Q^2 = (QR)^4 = R^2 = (RP)^2$

Fibrifold name	Couplings for γ P (I)	Point group	Intern. no.
$[4_0*{\cdot}2]$	$0+0+0-$	$*224$	127
$[4_2*{\cdot}2]$	$\frac{1}{2}+0+0-$	$*224$	136
$[4_0*{:}2]$	$0+\frac{1}{2}+0-$	$*224$	128
$[4_2*{:}2]$	$\frac{1}{2}+\frac{1}{2}+0-$	$*224$	135
$(4_0*{\cdot}2)$	$0+0+$	$*44$	100
$(4_1*{\cdot}2)$	$\frac{1}{4}+0+$	$*44$	109
$(4_2*{\cdot}2)$	$\frac{1}{2}+0+$	$*44$	102
$(4_0*{:}2)$	$0+\frac{1}{2}+$	$*44$	104
$(4_1*{:}2)$	$\frac{1}{4}+\frac{1}{2}+$	$*44$	110
$(4_2*{:}2)$	$\frac{1}{2}+\frac{1}{2}+$	$*44$	106
$(4\bar{\ast}{\cdot}2)$	$0-0+$	$2*2$	113
$(4\bar{\ast}{:}2)$	$0-\frac{1}{2}+$	$2*2$	114
(4_0*2_0)	$0+0-$	224	90
(4_1*2_1)	$\frac{1}{4}+0-$	224	92
(4_2*2_0)	$\frac{1}{2}+0-$	224	94
$(4\bar{\ast}_0 2_0)$	$0-0-$	$2*2$	117
$(4\bar{\ast}_1 2_0)$	$0-\frac{1}{2}-$	$2*2$	118
$(4\bar{\ast}2_1)$	$0-\frac{1}{4}-$	$2*2$	122

Table 25.4. Plane group: $4*2$; Relations $\gamma 4*{}^P 2$: $1 = \gamma^4 = P^2 = [P,\gamma]^2$

Fibrifold name	Couplings for γ δ ϵ (I)	Point group	Intern. no.
$[4_0 4_0 2_0]$	$0+0+0+0-$	$4*$	83
$[4_2 4_2 2_0]$	$\frac{1}{2}+\frac{1}{2}+0+0-$	$4*$	84
$[4_2 4_0 2_1]$	$\frac{1}{2}+0+\frac{1}{2}+0-$	$4*$	87
$(4_0 4_0 2_0)$	$0+0+0+$	44	75
$(4_1 4_1 2_1)$	$\frac{1}{4}+\frac{1}{4}+\frac{1}{2}+$	44	76
$(4_2 4_2 2_0)$	$\frac{1}{2}+\frac{1}{2}+0+$	44	77
$(4_2 4_0 2_1)$	$\frac{1}{2}+0+\frac{1}{2}+$	44	79
$(4_3 4_1 2_0)$	$\frac{3}{4}+\frac{1}{4}+0+$	44	80
$(4\,4_0 2)$	$0+0-0-$	$4*$	85
$(4\,4_1 2)$	$\frac{1}{4}+0-\frac{1}{4}-$	$4*$	88
$(4\,4_2 2)$	$\frac{1}{2}+0-\frac{1}{2}-$	$4*$	86
$(4\,4\,2_0)$	$0-0-0+$	$2\times$	81
$(4\,4\,2_1)$	$\frac{1}{2}-0-\frac{1}{2}+$	$2\times$	82

Table 25.5. Plane group: 442; Relations $\gamma 4\,{}^\delta 4\,{}^\epsilon 2$: $1 = \gamma^4 = \delta^4 = \epsilon^2 = \gamma\delta\epsilon$

Fibrifold name	Couplings for P Q R (I)	Point group	Intern. no.
$[*{\cdot}3{\cdot}3{\cdot}3]$	$0+0+0+0-$	$*223$	187
$[*{:}3{:}3{:}3]$	$\frac{1}{2}+\frac{1}{2}+\frac{1}{2}+0-$	$*223$	188
$(*{\cdot}3{\cdot}3{\cdot}3)$	$0+0+0+$	$*33$	156
$(*{:}3{:}3{:}3)$	$\frac{1}{2}+\frac{1}{2}+\frac{1}{2}+$	$*33$	158
$(\bar{\ast}3_0 3_0 3_0)$	$0-0-0-$	223	149
$(*3_1 3_1 3_1)$	$\frac{2}{3}-\frac{1}{3}-0-$	223	151
$(*3_0 3_1 3_2)$	$\frac{1}{3}-\frac{1}{3}-0-$	223	155

Table 25.6. Plane group: $*333$; Relations $*{}^P 3\,{}^Q 3\,{}^R 3$: $1 = P^2 = (PQ)^3 = Q^2 = (QR)^3 = R^2 = (RP)^3$

Fibrifold name	Couplings for γ P (I)	Point group	Intern. no.
$[3_0*{\cdot}3]$	$0+0+0-$	$*223$	189
$[3_0*{:}3]$	$0+\frac{1}{2}+0-$	$*223$	190
$(3_0*{\cdot}3)$	$0+0+$	$*33$	157
$(3_1*{\cdot}3)$	$\frac{1}{3}+0+$	$*33$	160
$(3_0*{:}3)$	$0+\frac{1}{2}+$	$*33$	159
$(3_1*{:}3)$	$\frac{1}{3}+\frac{1}{2}+$	$*33$	161
(3_0*3_0)	$0+0-$	223	150
(3_1*3_1)	$\frac{1}{3}+0-$	223	152

Table 25.7. Plane group: $3*3$; Relations $\gamma 3*{}^P 3$: $1 = \gamma^3 = P^2 = [P,\gamma]^3$

Fibrifold name	Couplings for γ δ ϵ (I)	Point group	Intern. no.
$[3_0 3_0 3_0]$	$0+0+0+0-$	$3*$	174
$(3_0 3_0 3_0)$	$0+0+0+$	33	143
$(3_1 3_1 3_1)$	$\frac{1}{3}+\frac{1}{3}+\frac{1}{3}+$	33	144
$(3_0 3_1 3_2)$	$0+\frac{1}{3}+\frac{2}{3}+$	33	146

Table 25.8. Plane group: 333; Relations $\gamma 3\,{}^\delta 3\,{}^\epsilon 3$: $1 = \gamma^3 = \delta^3 = \epsilon^3 = \gamma\delta\epsilon$

Fibrifold name	Couplings for $P\ Q\ R\ S\ (I)$	Point group	Intern. no.
$[*\!\cdot\!2\!\cdot\!2\!\cdot\!2\!\cdot\!2]$	$0+0+0+0+0-$	$*222$	47
$[*\!\cdot\!2\!\cdot\!2\!:\!2\!\cdot\!2]$	$0+0+0+\frac{1}{2}+0-$	$*222$	65
$[*\!\cdot\!2\!\cdot\!2\!:\!2\!:\!2]$	$0+0+\frac{1}{2}+\frac{1}{2}+0-$	$*222$	69
$[*\!\cdot\!2\!:\!2\!\cdot\!2\!:\!2]$	$0+\frac{1}{2}+0+\frac{1}{2}+0-$	$*222$	51
$[*\!\cdot\!2\!:\!2\!:\!2\!:\!2]$	$0+\frac{1}{2}+\frac{1}{2}+\frac{1}{2}+0-$	$*222$	67
$[*\!:\!2\!:\!2\!:\!2\!:\!2]$	$\frac{1}{2}+\frac{1}{2}+\frac{1}{2}+\frac{1}{2}+0-$	$*222$	49
$(*\!\cdot\!2\!\cdot\!2\!\cdot\!2\!\cdot\!2)$	$0+0+0+0+$	$*22$	25
$(*\!\cdot\!2\!\cdot\!2\!\cdot\!2\!:\!2)$	$0+0+0+\frac{1}{2}+$	$*22$	38
$(*\!\cdot\!2\!\cdot\!2\!:\!2\!:\!2)$	$0+0+\frac{1}{2}+\frac{1}{2}+$	$*22$	42
$(*\!\cdot\!2\!:\!2\!\cdot\!2\!:\!2)$	$0+\frac{1}{2}+0+\frac{1}{2}+$	$*22$	26
$(*\!\cdot\!2\!:\!2\!:\!2\!:\!2)$	$0+\frac{1}{2}+\frac{1}{2}+\frac{1}{2}+$	$*22$	39
$(*\!:\!2\!:\!2\!:\!2\!:\!2)$	$\frac{1}{2}+\frac{1}{2}+\frac{1}{2}+\frac{1}{2}+$	$*22$	27
$(*2\!\cdot\!2\!\cdot\!2\!\cdot\!2)$	$0+0+0+0-$	$*222$	51
$(*2\!\cdot\!2\!\cdot\!2\!:\!2)$	$0+0+\frac{1}{2}+0-$	$*222$	63
$(*2\!\cdot\!2\!:\!2\!\cdot\!2)$	$0+\frac{1}{2}+0+0-$	$*222$	55
$(*2\!\cdot\!2\!:\!2\!:\!2)$	$0+\frac{1}{2}+\frac{1}{2}+0-$	$*222$	64
$(*2\!:\!2\!\cdot\!2\!:\!2)$	$\frac{1}{2}+0+\frac{1}{2}+0-$	$*222$	57
$(*2\!:\!2\!:\!2\!:\!2)$	$\frac{1}{2}+\frac{1}{2}+\frac{1}{2}+0-$	$*222$	54
$(*2_0 2\!\cdot\!2\!\cdot\!2)$	$0-0-0+0+$	$*222$	67
$(*2_1 2\!\cdot\!2\!\cdot\!2)$	$\frac{1}{2}-0-0+0+$	$*222$	74
$(*2_0 2\!\cdot\!2\!:\!2)$	$0-0-0+\frac{1}{2}+$	$*222$	72
$(*2_1 2\!\cdot\!2\!:\!2)$	$\frac{1}{2}-0-0+\frac{1}{2}+$	$*222$	64
$(*2_0 2\!:\!2\!:\!2)$	$0-0-\frac{1}{2}+\frac{1}{2}+$	$*222$	68
$(*2_1 2\!:\!2\!:\!2)$	$\frac{1}{2}-0-\frac{1}{2}+\frac{1}{2}+$	$*222$	73
$(*2\!\cdot\!2\ 2\!\cdot\!2)$	$0-0+0-0+$	$2*$	10
$(*2\!\cdot\!2\ 2\!:\!2)$	$0-0+0-\frac{1}{2}+$	$2*$	12
$(*2\!:\!2\ 2\!:\!2)$	$0-\frac{1}{2}+0-\frac{1}{2}+$	$2*$	13
$(*2_0 2_0 2\!\cdot\!2)$	$0-0-0-0+$	$*222$	49
$(*2_0 2_1 2\!\cdot\!2)$	$0-0-\frac{1}{2}-0+$	$*222$	66
$(*2_1 2_1 2\!\cdot\!2)$	$\frac{1}{2}-0-\frac{1}{2}-0+$	$*222$	53
$(*2_0 2_0 2\!:\!2)$	$0-0-0-\frac{1}{2}+$	$*222$	50
$(*2_0 2_1 2\!:\!2)$	$0-0-\frac{1}{2}-\frac{1}{2}+$	$*222$	68
$(*2_1 2_1 2\!:\!2)$	$\frac{1}{2}-0-\frac{1}{2}-\frac{1}{2}+$	$*222$	54
$(*2_0 2_0 2_0 2_0)$	$0-0-0-0-$	222	16
$(*2_0 2_0 2_1 2_1)$	$0-0-0-\frac{1}{2}-$	222	21
$(*2_0 2_1 2_0 2_1)$	$0-0-\frac{1}{2}-\frac{1}{2}-$	222	22
$(*2_1 2_1 2_1 2_1)$	$0-\frac{1}{2}-0-\frac{1}{2}-$	222	17

Fibrifold name	Couplings for $\gamma\ P\ Q\ (I)$	Point group	Intern. no.
$[2_0*\!\cdot\!2\!\cdot\!2]$	$0+0+0+0-$	$*222$	65
$[2_1*\!\cdot\!2\!\cdot\!2]$	$\frac{1}{2}+0+0+0-$	$*222$	71
$[2_0*\!\cdot\!2\!:\!2]$	$0+0+\frac{1}{2}+0-$	$*222$	74
$[2_1*\!\cdot\!2\!:\!2]$	$\frac{1}{2}+0+\frac{1}{2}+0-$	$*222$	63
$[2_0*\!:\!2\!:\!2]$	$0+\frac{1}{2}+\frac{1}{2}+0-$	$*222$	66
$[2_1*\!:\!2\!:\!2]$	$\frac{1}{2}+\frac{1}{2}+\frac{1}{2}+0-$	$*222$	72
$(2_0*\!\cdot\!2\!\cdot\!2)$	$0+0+0+$	$*22$	35
$(2_1*\!\cdot\!2\!\cdot\!2)$	$\frac{1}{2}+0+0+$	$*22$	44
$(2_0*\!\cdot\!2\!:\!2)$	$0+0+\frac{1}{2}+$	$*22$	46
$(2_1*\!\cdot\!2\!:\!2)$	$\frac{1}{2}+0+\frac{1}{2}+$	$*22$	36
$(2_0*\!:\!2\!:\!2)$	$0+\frac{1}{2}+\frac{1}{2}+$	$*22$	37
$(2_1*\!:\!2\!:\!2)$	$\frac{1}{2}+\frac{1}{2}+\frac{1}{2}+$	$*22$	45
$(2_0*2\!\cdot\!2)$	$0+0+0-$	$*222$	53
$(2_1*2\!\cdot\!2)$	$\frac{1}{2}+0+0-$	$*222$	58
$(2_0*2\!:\!2)$	$0+\frac{1}{2}+0-$	$*222$	52
$(2_1*2\!:\!2)$	$\frac{1}{2}+\frac{1}{2}+0-$	$*222$	60
$(2\bar{*}\!\cdot\!2\!\cdot\!2)$	$0-0+0+$	$*222$	59
$(2\bar{*}\!\cdot\!2\!:\!2)$	$0-0+\frac{1}{2}+$	$*222$	62
$(2\bar{*}\!:\!2\!:\!2)$	$0-\frac{1}{2}+\frac{1}{2}+$	$*222$	56
$(2\bar{*}2\!\cdot\!2)$	$0-0-0+$	$2*$	12
$(2\bar{*}2\!:\!2)$	$0-0-\frac{1}{2}+$	$2*$	15
$(2_0*2_0 2_0)$	$0+0-0-$	222	21
$(2_1*2_0 2_0)$	$\frac{1}{2}+0-0-$	222	23
$(2_0*2_1 2_1)$	$0+\frac{1}{2}-0-$	222	24
$(2_1*2_1 2_1)$	$\frac{1}{2}+\frac{1}{2}-0-$	222	20
$(2\bar{*}_0 2_0 2_0)$	$0-0-0-$	$*222$	50
$(2\bar{*}_1 2_0 2_0)$	$0-\frac{1}{2}-\frac{1}{2}-$	$*222$	48
$(2\bar{*}2_0 2_1)$	$0-\frac{1}{4}-\frac{1}{4}-$	$*222$	70
$(2\bar{*}2_1 2_1)$	$0-0-\frac{1}{2}-$	$*222$	52

Table 25.10. Plane group: $2*22$; Relations $\gamma 2*^{P_2 Q_2}: 1 = \gamma^2 = P^2 = (PQ)^2 = Q^2 = (Q\gamma P\gamma^{-1})^2$

Table 25.9. Plane group: $*2222$; Relations $*^{P_2 Q_2 R_2 S_2}$: $1 = P^2 = (PQ)^2 = Q^2 = (QR)^2 = R^2 = (RS)^2 = S^2 = (SP)^2$

Fibrifold name	Couplings for γ δ P (I)	Point group	Intern. no.
$[2_0 2_0 *\cdot]$	$0+0+0+0-$	$*222$	51
$[2_0 2_1 *\cdot]$	$0+\frac{1}{2}+0+0-$	$*222$	63
$[2_1 2_1 *\cdot]$	$\frac{1}{2}+\frac{1}{2}+0+0-$	$*222$	59
$[2_0 2_0 *{:}]$	$0+0+\frac{1}{2}+0-$	$*222$	53
$[2_0 2_1 *{:}]$	$0+\frac{1}{2}+\frac{1}{2}+0-$	$*222$	64
$[2_1 2_1 *{:}]$	$\frac{1}{2}+\frac{1}{2}+\frac{1}{2}+0-$	$*222$	57
$(2_0 2_0 *\cdot)$	$0+0+0+$	$*22$	28
$(2_0 2_1 *\cdot)$	$0+\frac{1}{2}+0+$	$*22$	40
$(2_1 2_1 *\cdot)$	$\frac{1}{2}+\frac{1}{2}+0+$	$*22$	31
$(2_0 2_0 *{:})$	$0+0+\frac{1}{2}+$	$*22$	30
$(2_0 2_1 *{:})$	$0+\frac{1}{2}+\frac{1}{2}+$	$*22$	41
$(2_1 2_1 *{:})$	$\frac{1}{2}+\frac{1}{2}+\frac{1}{2}+$	$*22$	29
$(2_0 2_0 *)$	$0+0+0-$	222	17
$(2_0 2_1 *)$	$0+\frac{1}{2}+0-$	222	20
$(2_1 2_1 *)$	$\frac{1}{2}+\frac{1}{2}+0-$	222	18
$(2_0 2̄ *\cdot)$	$0+0-0+$	$*222$	57
$(2_1 2̄ *\cdot)$	$\frac{1}{2}+0-0+$	$*222$	62
$(2_0 2̄ *{:})$	$0+0-\frac{1}{2}+$	$*222$	60
$(2_1 2̄ *{:})$	$\frac{1}{2}+0-\frac{1}{2}+$	$*222$	61
$(2_0 2̄ *_0)$	$0+0-0-$	$*222$	54
$(2_0 2̄ *_1)$	$0+\frac{1}{2}-0-$	$*222$	52
$(2_1 2̄ *_0)$	$\frac{1}{2}+\frac{1}{2}-0-$	$*222$	56
$(2_1 2̄ *_1)$	$\frac{1}{2}+0-0-$	$*222$	60
$(2\,2 *\cdot)$	$0-0-0+$	$2*$	11
$(2\,2 *{:})$	$0-0-\frac{1}{2}+$	$2*$	14
$(2\,2 *_0)$	$0-0-0-$	$2*$	13
$(2\,2 *_1)$	$0-\frac{1}{2}-0-$	$2*$	15

Table 25.11. Plane group: $22*$; Relations $^{\gamma}2\,^{\delta}2*^{P}$: $1 = \gamma^2 = \delta^2 = P^2 = [P, \gamma\delta]$

Fibrifold name	Couplings for γ δ Z (I)	Point group	Intern. no.
$[2_0 2_0 \times_0]$	$0+0+0+0-$	$*222$	55
$[2_0 2_0 \times_1]$	$0+0+\frac{1}{2}+0-$	$*222$	58
$[2_1 2_1 \times]$	$\frac{1}{2}+\frac{1}{2}+0+0-$	$*222$	62
$(2_0 2_0 \times_0)$	$0+0+0+$	$*22$	32
$(2_0 2_0 \times_1)$	$0+0+\frac{1}{2}+$	$*22$	34
$(2_0 ?_1 \times)$	$0+\frac{1}{2}+\frac{1}{4}+$	$*22$	43
$(2_1 2_1 \times)$	$\frac{1}{2}+\frac{1}{2}+0+$	$*22$	33
$(2_0 2_0 \bar{\times})$	$0+0+0-$	222	18
$(2_1 2_1 \bar{\times})$	$\frac{1}{2}+\frac{1}{2}+0-$	222	19
$(2\,2 \times)$	$0-0-0+$	$2*$	14

Table 25.12. Plane group: $22\times$; Relations $^{\gamma}2\,^{\delta}2\times^{Z}$: $1 = \gamma^2 = \delta^2 = \gamma\delta Z^2$

Fibrifold name	Couplings for γ δ ϵ ζ (I)	Point group	Intern. no.
$[2_0 2_0 2_0 2_0]$	$0+0+0+0+0-$	$2*$	10
$[2_0 2_0 2_1 2_1]$	$0+0+\frac{1}{2}+\frac{1}{2}+0-$	$2*$	12
$[2_1 2_1 2_1 2_1]$	$\frac{1}{2}+\frac{1}{2}+\frac{1}{2}+\frac{1}{2}+0-$	$2*$	11
$(2_0 2_0 2_0 2_0)$	$0+0+0+0+$	22	3
$(2_0 2_0 2_1 2_1)$	$0+0+\frac{1}{2}+\frac{1}{2}+$	22	5
$(2_1 2_1 2_1 2_1)$	$\frac{1}{2}+\frac{1}{2}+\frac{1}{2}+\frac{1}{2}+$	22	4
$(2_0 2_0 2\,2)$	$0+0+0-0-$	$2*$	13
$(2_0 2_1 2\,2)$	$0+\frac{1}{2}+0-\frac{1}{2}-$	$2*$	15
$(2_1 2_1 2\,2)$	$\frac{1}{2}+\frac{1}{2}+0-0-$	$2*$	14
$(2\,2\,2\,2)$	$0-0-0-0-$	$\times$	2

Table 25.13. Plane group: 2222; Relations $^{\gamma}2\,^{\delta}2\,^{\epsilon}2\,^{\zeta}2$: $1 = \gamma^2 = \delta^2 = \epsilon^2 = \zeta^2 = \gamma\delta\epsilon\zeta$

Fibrifold name	Couplings for λ P Q (I)	Point group	Intern. no.
[*₀·*₀·]	$0+0+0+0-$	*22	25
[*₁·*₁·]	$\frac{1}{2}+0+0+0-$	*22	38
[*₀·*₀:]	$0+0+\frac{1}{2}+0-$	*22	35
[*₁·*₁:]	$\frac{1}{2}+0+\frac{1}{2}+0-$	*22	42
[*₀:*₀:]	$0+\frac{1}{2}+\frac{1}{2}+0-$	*22	28
[*₁:*₁:]	$\frac{1}{2}+\frac{1}{2}+\frac{1}{2}+0-$	*22	39
(*·*·)	$0+0+0+$	*	6
(*·*:)	$0+0+\frac{1}{2}+$	*	8
(*:*:)	$0+\frac{1}{2}+\frac{1}{2}+$	*	7
(*̄·*̄·)	$0-0+0+$	*22	26
(*̄·*̄:)	$0-0+\frac{1}{2}+$	*22	36
(*̄:*̄:)	$0-\frac{1}{2}+\frac{1}{2}+$	*22	29
(*·*₀)	$0+0+0-$	*22	28
(*·*₁)	$\frac{1}{2}+0+0-$	*22	40
(*:*₀)	$0+\frac{1}{2}+0-$	*22	32
(*:*₁)	$\frac{1}{2}+\frac{1}{2}+0-$	*22	41
(*̄·*̄₀)	$0-0+0-$	*22	39
(*̄·*̄₁)	$\frac{1}{2}-0+0-$	*22	46
(*̄:*̄₀)	$0-\frac{1}{2}+0-$	*22	45
(*̄:*̄₁)	$\frac{1}{2}-\frac{1}{2}+0-$	*22	41
(*₀*₀)	$0+0-0-$	22	3
(*₁*₁)	$\frac{1}{2}+0-0-$	22	5
(*̄₀*̄₀)	$0-0-0-$	*22	27
(*̄₀*̄₁)	$0-0-\frac{1}{2}-$	*22	37
(*̄₁*̄₁)	$0-\frac{1}{2}-\frac{1}{2}-$	*22	30

Table 25.14. Plane group: **; Relations $^{\lambda}*^{P}*^{Q}$: $1 = P^2 = Q^2 = [\lambda, P] = [\lambda, Q]$

Fibrifold name	Couplings for P Z (I)	Point group	Intern. no.
[*·×₀]	$0+0+0-$	*22	38
[*·×₁]	$0+\frac{1}{2}+0-$	*22	44
[*:×₀]	$\frac{1}{2}+0+0-$	*22	46
[*:×₁]	$\frac{1}{2}+\frac{1}{2}+0-$	*22	40
(*·×)	$0+0+$	*	8
(*:×)	$\frac{1}{2}+0+$	*	9
(*·×̄)	$0+0-$	*22	31
(*:×̄)	$\frac{1}{2}+0-$	*22	33
(*₀×₀)	$0-0+$	*22	30
(*₀×₁)	$0-\frac{1}{2}+$	*22	34
(*₁×)	$0-\frac{1}{4}+$	*22	43
(*×̄)	$0-0-$	22	5

Table 25.15. Plane group: *×; Relations $*^{P}×^{Z}$: $1 = P^2 = [P, Z^2]$

Fibrifold name	Couplings for Y Z (I)	Point group	Intern. no.
[×₀×₀]	$0+0+0-$	*22	26
[×₀×₁]	$0+\frac{1}{2}+0-$	*22	36
[×₁×₁]	$\frac{1}{2}+\frac{1}{2}+0-$	*22	31
(××₀)	$0+0+$	*	7
(××₁)	$0+\frac{1}{2}+$	*	9
(×̄×₀)	$0-0+$	*22	29
(×̄×₁)	$0-\frac{1}{2}+$	*22	33
(×̄×̄)	$0-0-$	22	4

Table 25.16. Plane group: ××; Relations $×^{Y}×^{Z}$: $1 = Y^2 Z^2$

Fibrifold name	Couplings for X Y (I)	Point group	Intern. no.
[○₀]	$0+0+0-$	*	6
[○₁]	$\frac{1}{2}+\frac{1}{2}+0-$	*	8
(○)	$0+0+$	1	1
(ō₀)	$0-0-$	*	7
(ō₁)	$0-\frac{1}{2}-$	*	9

Table 25.17. Plane group: ○; Relations $○^{X Y}$: $1 = [X, Y]$

Examples and Exercises

Try your hand at verifying the couplings in the following pictures!

1. $(*\cdot 6\,3_1\,2)$ with coupling $0+\frac{1}{3}-0-$.

2. $(*.6.3.2)$, with coupling $0 + 0 + 0+$

3. $(*6_2 3_2 2_0)$ with coupling $0 - \frac{2}{3} - 0 -$.

4. $(*{:}6{\cdot}3{\cdot}2)$ with coupling $\frac{1}{2} + 0 + 0 +$.

5. $(*{:}6\,3_1 2)$ with coupling $\frac{1}{2} + \frac{1}{3} - 0 -$.

6. $(*6_1 3_1 2_1)$ with coupling $\frac{1}{2} - \frac{1}{3} - 0 -$.

7. $(*6{:}3{:}2)$ with coupling $0 - \frac{1}{2} + \frac{1}{2} +$.

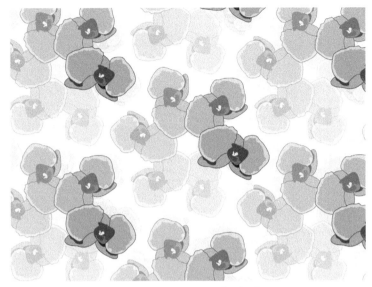

8. $(6_1 3_1 2_1)$ with coupling $\frac{1}{6} + \frac{1}{3} + \frac{1}{2} +$.

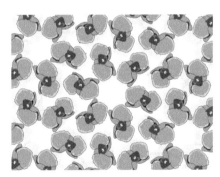

9. $(6_0 3_0 2_0)$ with coupling $0 + 0 + 0 +$.

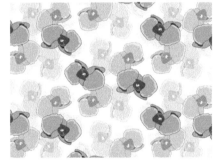

10. $(6_2 3_2 2_0)$ with coupling $\frac{1}{3} + \frac{2}{3} + 0 +$.

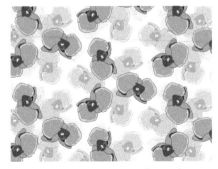

11. $(6_3 3_0 2_1)$ with coupling $\frac{1}{2} + 0 + \frac{1}{2} +$.

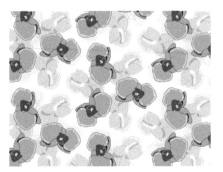

12. $(6\, 3_0\, 2)$ with coupling $0 - 0 + 0 -$.

13. $(4_1*{:}2)$ with coupling $\frac{1}{4}+\frac{1}{2}+$.

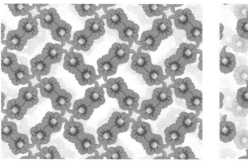

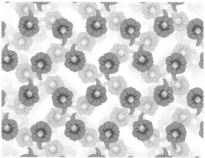

14. $(4_0*{\cdot}2)$ with coupling $0+0+$

15. $(4_2*{:}2)$ with coupling $\frac{1}{2}+\frac{1}{2}+$.

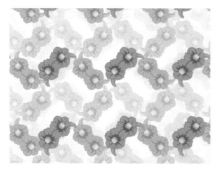

16. $(4_1*{\cdot}2)$ with coupling $\frac{1}{4}+0+$.

17. $(4\bar{*}{:}2)$ with coupling $0-\frac{1}{2}+$.

18. $(\bar{\times} \times_1)$ with coupling $0 - \frac{1}{2} +$.

19. $(\bar{\times} \bar{\times})$ with coupling $0 - 0 -$.

20. $(\bar{\times} \times_0.)$ with coupling $0 - 0 +$

21. $[\times_0 \times_0]$ with coupling $0 + 0 + 0 -$.

22. $(\times \times_1)$ with coupling $0 + \frac{1}{2} +$.

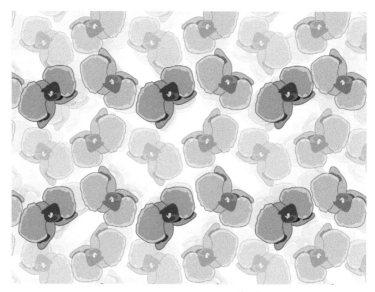

23. $(2_0 2_1 \times)$ with coupling $0 + \frac{1}{2} + \frac{1}{4} +$.

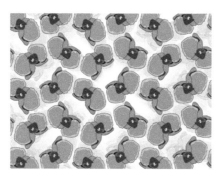

24. $(2_0 2_0 \times_0)$ with coupling $0 + 0 + 0 +$.

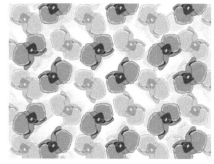

25. $(2_0 2_0 \times_1)$ with coupling $0 + 0 + \frac{1}{2} +$.

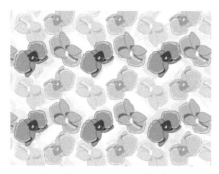

26. $(2_1 2_1 \bar{\times})$ with coupling $\frac{1}{2} + \frac{1}{2} + 0 -$.

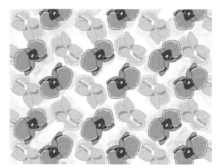

27. $(2_0 2_0 \bar{\times})$ with coupling $0 + 0 + 0 -$.

– 26 –

Higher Still

Four-Dimensional Point Groups

There are only a few types of higher dimensional group for which we have complete enumerations. Next in sequence are the four-dimensional point groups, which of course can also be regarded as three-dimensional spherical groups, since they act on the unit sphere $x^2+y^2+z^2+t^2 = 1$. Their enumeration was started by Goursat [16], continued by Seifert and Threlfall [28], and also by Du Val [29]—we shall comment on their roles later.

The groups are usually described in terms of quaternions, because the general orthogonal operation has the form

$$[q,r] : x \rightarrow q^{-1}xr \text{ or}$$
$$*[q,r] : x \rightarrow q^{-1}\bar{x}r$$

accordingly as its determinant is plus or minus 1.

We should only briefly explain Tables 26.1, 26.2, and 26.3, which are quoted from *On Quaternions and Octonians* [26]. The name of the typical group has a form $\pm\frac{1}{f}[\mathbf{L} \times \mathbf{R}]$ if it contains -1 (the negative of the identity matrix), so that its elements come in pairs $\pm g$, and otherwise $+\frac{1}{f}[\mathbf{L}\times\mathbf{R}]$. These names are appropriate because, up to sign, the group contains the proportion $\frac{1}{f}$ of the elements of a direct product of the groups $\mathbf{L}$ and $\mathbf{R}$ that consist of left and right multiplications, respectively.

A generator $[l,r]$ denotes the map $x \rightarrow \bar{l}xr$ while the names of individual quaternions are given below Table 26.1. When only a $+$ or $-$ is given, it is to be applied to a generator written explicitly

---

Group	Generators
$\pm[I \times O]$	$[i_I, 1], [\omega, 1], [1, i_O], [1, \omega];$
$\pm[I \times T]$	$[i_I, 1], [\omega, 1], [1, i], [1, \omega];$
$\pm[I \times D_{2n}]$	$[i_I, 1], [\omega, 1], [1, e_n], [1, j];$
$\pm[I \times C_n]$	$[i_I, 1], [\omega, 1], [1, e_n];$
$\pm[O \times T]$	$[i_O, 1], [\omega, 1], [1, i], [1, \omega];$
$\pm[O \times D_{2n}]$	$[i_O, 1], [\omega, 1], [1, e_n], [1, j];$
$\pm\frac{1}{2}[O \times D_{2n}]$	$[i, 1], [\omega, 1], [1, e_n]; [i_O, j]$
$\pm\frac{1}{2}[O \times \overline{D}_{4n}]$	$[i, 1], [\omega, 1], [1, e_n], [1, j]; [i_O, e_{2n}]$
$\pm\frac{1}{6}[O \times D_{6n}]$	$[i, 1], [j, 1], [1, e_n]; [i_O, j], [\omega, e_{3n}]$
$\pm[O \times C_n]$	$[i_O, 1], [\omega, 1], [1, e_n];$
$\pm\frac{1}{2}[O \times C_{2n}]$	$[i, 1], [\omega, 1], [1, e_n]; [i_O, e_{2n}]$
$\pm[T \times D_{2n}]$	$[i, 1], [\omega, 1], [1, e_n], [1, j];$
$\pm[T \times C_n]$	$[i, 1], [\omega, 1], [1, e_n];$
$\pm\frac{1}{3}[T \times C_{3n}]$	$[i, 1], [1, e_n]; [\omega, e_{3n}]$
$\pm\frac{1}{2}[D_{2m} \times \overline{D}_{4n}]$	$[e_m, 1], [1, e_n], [1, j]; [j, e_{2n}]$
$\pm[D_{2m} \times C_n]$	$[e_m, 1], [j, 1], [1, e_n];$
$\pm\frac{1}{2}[D_{2m} \times C_{2n}]$	$[e_m, 1], [1, e_n]; [j, e_{2n}]$
$+\frac{1}{2}[D_{2m} \times C_{2n}]$	$-\quad,\quad -\quad;\quad +$
$\pm\frac{1}{2}[\overline{D}_{4m} \times C_{2n}]$	$[e_m, 1], [j, 1], [1, e_n]; [e_{2m}, e_{2n}]$

Key:
$$i_I = \frac{i + \sigma j + \tau k}{2} \ (\sigma = \frac{\sqrt{5}-1}{2}, \tau = \frac{\sqrt{5}+1}{2}),$$

$$i_O = \frac{j + k}{\sqrt{2}}, \omega = \frac{-1 + i + j + k}{2}, i_T = i, e_n = e^{\pi i/n}$$

Table 26.1. Chiral groups, I. These are most of the "metachiral" groups—some others appear in the last few lines of Table 26.2.

in a previous line. Generators before a ";" are pure left or right multiplications; those after involve both.

We now comment on the history of the enumerations. What Goursat enumerated, strictly speaking, were the groups of motions in elliptic three-space. This is equivalent to finding the subgroups of SO_4 that contain -1. He missed the group $\pm\frac{1}{4}(D_{4m} \times D_{4n})$, as was noted by Seifert and Threlfall and independently by J. Sunday. Seifert and Threlfall extended this to the general case and

Group	Generators	Coxeter Name
$\pm[I \times I]$	$[i_I, 1], [\omega, 1], [1, i_I], [1, \omega];$	$[3, 3, 5]^+$
$\pm\frac{1}{60}[I \times I]$	$;[\omega, \omega], [i_I, i_I]$	$2.[3, 5]^+$
$+\frac{1}{60}[I \times I]$	$; \ + \ , \ +$	$[3, 5]^+$
$\pm\frac{1}{60}[I \times \overline{I}]$	$;[\omega, \omega], [i_I, i'_I]$	$2.[3, 3, 3]^+$
$+\frac{1}{60}[I \times \overline{I}]$	$; \ + \ , \ +$	$[3, 3, 3]^+$
$\pm[O \times O]$	$[i_O, 1], [\omega, 1], [1, i_O], [1, \omega];$	$[3, 4, 3]^+ : 2$
$\pm\frac{1}{2}[O \times O]$	$[i, 1], [\omega, 1], [1, i], [1, \omega];[i_O, i_O]$	$[3, 4, 3]^+$
$\pm\frac{1}{6}[O \times O]$	$[i, 1], [j, 1], [1, i], [1, j];[\omega, \omega], [i_O, i_O]$	$[3, 3, 4]^+$
$\pm\frac{1}{24}[O \times O]$	$;[\omega, \omega], [i_O, i_O]$	$2.[3, 4]^+$
$+\frac{1}{24}[O \times O]$	$; \ + \ , \ +$	$[3, 4]^+$
$+\frac{1}{24}[O \times \overline{O}]$	$; \ + \ , \ -$	$[2, 3, 3]^+$
$\pm[T \times T]$	$[i, 1], [\omega, 1], [1, i], [1, \omega];$	$[^+3, 4, 3^+]$
$\pm\frac{1}{3}[T \times T]$	$[i, 1], [j, 1], [1, i], [1, j];[\omega, \omega]$	$[^+3, 3, 4^+]$
$\cong \pm\frac{1}{3}[T \times \overline{T}]$	$[i, 1], [j, 1], [1, i], [1, j];[\omega, \overline{\omega}]$	"
$\pm\frac{1}{12}[T \times T]$	$;[\omega, \omega], [i, i]$	$2.[3, 3]^+$
$\cong \pm\frac{1}{12}[T \times \overline{T}]$	$;[\omega, \overline{\omega}], [i, -i]$	"
$+\frac{1}{12}[T \times T]$	$; \ + \ , \ +$	$[3, 3]^+$
$\cong +\frac{1}{12}[T \times \overline{T}]$	$; \ + \ , \ +$	"
$\pm[D_{2m} \times D_{2n}]$	$[e_m, 1], [j, 1], [1, e_n], [1, j];$	
$\pm\frac{1}{2}[\overline{D}_{4m} \times \overline{D}_{4n}]$	$[e_m, 1], [j, 1], [1, e_n], [1, j];[e_{2m}, e_{2n}]$	
$\pm\frac{1}{4}[D_{4m} \times \overline{D}_{4n}]$	$[e_m, 1], [1, e_n];[e_{2m}, j], [j, e_{2n}]$	Conditions
$+\frac{1}{4}[D_{4m} \times \overline{D}_{4n}]$	$- \ , \ - \ ; \ + \ , \ +$	m, n odd
$\pm\frac{1}{2f}[D_{2mf} \times D_{2nf}^{(s)}]$	$[e_m, 1], [1, e_n];[e_{mf}, e_{nf}^s], [j, j]$	$(s, f) = 1$
$+\frac{1}{2f}[D_{2mf} \times D_{2nf}^{(s)}]$	$- \ , \ - \ ; \ + \ , \ +$	m, n odd, $(s, 2f) = 1$
$\pm\frac{1}{f}[C_{mf} \times C_{nf}^{(s)}]$	$[e_m, 1], [1, e_n];[e_{mf}, e_{nf}^s]$	$(s, f) = 1$
$+\frac{1}{f}[C_{mf} \times C_{nf}^{(s)}]$	$- \ , \ - \ ; \ +$	m, n odd, $(s, 2f) = 1$

Table 26.2. Chiral groups, II. These groups are mostly "orthochiral," with a few "parachiral" groups in the last few lines. The generators should be taken with both signs except in the haploid cases, for which we just indicate the proper choice of sign.

Group	Extending element	Coxeter Name
$\pm[I \times I] \cdot 2$	$*$	$[3,3,5]$
$\pm\frac{1}{60}[I \times I] \cdot 2$	$*$	$2.[3,5]$
$+\frac{1}{60}[I \times I] \cdot 2_{3 \, or \, 1} 2_3$ or 2_1	$*$ or $-*$	$[3,5]$ or $[3,5]^\circ$
$\pm\frac{1}{60}[I \times \bar{I}] \cdot 2$	$*$	$2.[3,3,3]$
$+\frac{1}{60}[I \times \bar{I}] \cdot 2_3$ or 2_1	$*$ or $-*$	$[3,3,3]^\circ$ or $[3,3,3]$
$\pm[O \times O] \cdot 2$	$*$	$[3,4,3]:2$
$\pm\frac{1}{2}[O \times O] \cdot 2$ or $\bar{2}$	$*$ or $*[1, i_O]$	$[3,4,3]$ or $[3,4,3]^{+\cdot}2$
$\pm\frac{1}{6}[O \times O] \cdot 2$	$*$	$[3,3,4]$
$\pm\frac{1}{24}[O \times O] \cdot 2$	$*$	$2.[3,4]$
$+\frac{1}{24}[O \times O] \cdot 2_3$ or 2_1	$*$ or $-*$	$[3,4]$ or $[3,4]^\circ$
$+\frac{1}{24}[O \times \bar{O}] \cdot 2_3$ or 2_1	$*$ or $-*$	$[2,3,3]^\circ$ or $[2,3,3]$
$\pm[T \times T] \cdot 2$	$*$	$[3,4,3^+]$
$\pm\frac{1}{3}[T \times T] \cdot 2$	$*$	$[^+3,3,4]$
$\pm\frac{1}{3}[T \times \bar{T}] \cdot 2$	$*$	$[3,3,4^+]$
$\pm\frac{1}{12}[T \times T] \cdot 2$	$*$	$2.[^+3,4]$
$\pm\frac{1}{12}[T \times \bar{T}] \cdot 2$	$*$	$2.[3,3]$
$+\frac{1}{12}[T \times T] \cdot 2_3$ or 2_1	$*$ or $-*$	$[^+3,4]$ or $[^+3,4]^\circ$
$+\frac{1}{12}[T \times \bar{T}] \cdot 2_3$ or 2_1	$*$ or $-*$	$[3,3]^\circ$ or $[3,3]$
$\pm[D_{2n} \times D_{2n}] \cdot 2$	$*$	
$\pm\frac{1}{2}[\overline{D}_{4n} \times \overline{D}_{4n}] \cdot 2$ or $\bar{2}$	$*$ or $*[1, e_{2n}]$	
$\pm\frac{1}{4}[D_{4n} \times \overline{D}_{4n}] \cdot 2$	$*$	<u>Conditions</u>
$+\frac{1}{4}[D_{4n} \times \overline{D}_{4n}] \cdot 2_3$ or 2_1	$*$ or $-*$	n odd
$\pm\frac{1}{2f}[D_{2nf} \times D_{2nf}^{(s)}] \cdot 2^{(\alpha,\beta)}$ or $\bar{2}$	$*[e_{2nf}^\alpha, e_{2nf}^{\alpha s+\beta f}]$ or $*[1,j]$	
$+\frac{1}{2f}[D_{2nf} \times D_{2nf}^{(s)}] \cdot 2^{(\alpha,\beta)}$ or $\bar{2}$	$*[e_{2nf}^\alpha, e_{2nf}^{\alpha s+\beta f}]$ or $*[1,j]$	See [23,
$\pm\frac{1}{f}[C_{nf} \times C_{nf}^{(s)}] \cdot 2^{(\gamma)}$	$*[1, e_{2nf}^{\gamma(f,s+1)}]$	pages 50–53]
$+\frac{1}{f}[C_{nf} \times C_{nf}^{(s)}] \cdot 2^{(\gamma)}$	$*[1, e_{2nf}^{\gamma(f,s+1)}]$	

Table 26.3. Achiral groups.

studied the orbifolds (as they are now known) of some of the groups. Du Val's elegant book on *Quaternions and Homographies* also contains an enumeration, but sadly omits the relatives of the group $\pm\frac{1}{4}(D_{4m}, D_{4n})$ that Goursat missed. There is a minor respect in which all enumerations before that of [26] were incomplete—namely some of the groups depend on several parameters and the exact conditions on the parameters for two groups to be isomorphic were first given in [26].

The four-dimensional space groups have also been enumerated. However, their enumeration takes up an entire book [2] and so we cannot summarize it here.

Higher-dimensional point groups have also been listed, but in more and more abbreviated ways as the dimension increases.

Regular Polytopes

The term *regular polytope* is usually understood to mean *"convex regular polytope."* These are

- in one dimension, the line segment with Schäfli symbol $\{\}$;

- in two dimensions, the regular polygons with symbols $\{3\}$, $\{4\}$, $\{5\}$, ...;

- in three dimensions, $\{3,3\}, \{3,4\}, \{3,5\}, \{4,3\}, \{5,3\}$;

- in four dimensions,

    - $\{3,3,3\}$ the simplex, or 5-cell,
    - $\{4,3,3\}$ the tesseract, or 8-cell,
    - $\{3,3,4\}$ the orthoplex, or 16-cell,
    - $\{3,4,3\}$ the polyoctahedron, or 24-cell,
    - $\{5,3,3\}$ the polydodecahedron, or 120-cell,
    - $\{3,3,5\}$ the polytetrahedron, or 600-cell;

- in $n \geq 5$ dimensions,

    - $\{3,3,\ldots,3,3\}$ the n-simplex,
    - $\{4,3,\ldots,3,3\}$ the n-hypercube,
    - $\{3,3,\ldots 3,4\}$ the n-orthoplex.

A gratuitous but pretty figure of a 5-coloring of the polydodecahedron.

Since lavish treatments are available elsewhere, we won't say much more about these here. But we will comment briefly on our proposed names, first for the polytopes that exist in all dimensions.

The well-established word "simplex" for $\{3, 3, \ldots 3, 3\}$ was apparently first used by Schoute in 1902, while "hypercube" for $\{4, 3, \ldots, 3, 3\}$ appears in a 1909 *Scientific American*. We adopt Conway and Sloan's "orthoplex" (abbreviating "orthant complex") for $\{3, 3, \ldots 3, 4\}$, which is appropriate since this has one cell for each orthant.

In the generic name "polytope" (introduced by Alicia Boole Stott), the suffix "-tope" refers to the shape of the cells. Since the cells of $\{5, 3, 3\}$, for example, are dodecahedra, this polytope can correctly be called the "polydodecahedron," and similarly $\{3, 4, 3\}$ the "polyoctahedron."

There are, of course, three four-dimensional polytopes with tetrahedral cells, $\{3, 3, 3\}$, $\{3, 3, 4\}$, and $\{3, 3, 5\}$; but since the first two have already been named, and "poly-" really means "many," we legitimately reserve the name "polytetrahedron" for $\{3, 3, 5\}$, which has no fewer than six hundred tetrahedral cells! The "poly-" terminology has the great advantage of extending nicely to the regular star polytopes discussed later.

Finally we remark that "hypercube" is often used also in four dimensions, but we have adopted Hinton's 1888 term "tesseract," which refers only to the four-dimensional case.

Other terms that have been used are "pentatope" for $\{3,3,3\}$, "measure-polytope" for $\{4,3,\ldots,3,3\}$, and "cross-polytope" for $\{3,3,\ldots,3,4\}$.

Four-Dimensional Archimedean Polytopes

The four-dimensional analogs of the Archimedean polyhedra were enumerated by Conway and Guy [7]. Since they only published a brief announcement, we shall describe the polytopes here. Almost all of them can be obtained by Wythoff's construction from the four-dimensional reflection groups, the irreducible ones having Coxeter diagrams:

The exceptions are the prisms on the antiprisms, snub cube and snub dodecahedron, which we don't illustrate, together with two very interesting ones we do. The latter are the "semi-snub polyoctahedron," discovered in 1900 by Thorold Gosset [15], and the "grand antiprism," discovered by Conway and Guy in 1965 [7].

The Coxeter diagram for a reflection group has a node for each mirror of the kaleidoscope that outlines its fundamental region. If a pair of mirrors meets at angle π/n, $n \geq 4$, the corresponding nodes are connected by a line marked n; we omit the marking for $n = 3$ and omit the line for $n = 2$.

The vertices of the typical polytope produced from a Coxeter diagram by Wythoff's construction are just the images of a suitably chosen point, which is sufficiently well indicated by ringing just those nodes corresponding to mirrors not containing the point.

There is an easier description when the reflection group is that of an n-dimensional regular polytope P, when the nodes can be numbered $0, 1, 2, \ldots, n-1$. Then if we ring only node i, we obtain the polytope i-ambo P, whose vertices are just the centers of the i-dimensional cells of P (so that "0-ambo P" is P itself). If instead we ring two nodes i and j, we obtain ij-ambo P, whose typical vertex is a suitably chosen point on the line-segment joining the centers of incident i-dimensional and j-dimensional cells of P. In general, the typical vertex of the n-dimensional polytope $ij\ldots k$-ambo P is a suitably chosen interior point of a simplex whose vertices are the centers of mutually incident i-dimensional, j-dimensional, $\ldots$, k-dimensional cells of P.

Such polytopes usually have two names, because if Q is the regular polyhedron dual to P, then $ij\ldots k$-ambo $P = IJ\ldots K$-ambo Q, where $i + I = j + J = \ldots = k + K = n - 1$. We have systematically chosen the name with the "lower" ambo-numbers.

This is not the only kind of repetition that occurs—there are certain cases in which extra symmetries can be adjoined. For example, each polytope P obtained by Wythoff's construction from the only finite four-dimensional reflection group that is not that of a regular polytope can also be obtained from one that is. For the group is

and we can always arrange that the upper and lower nodes of this are either both ringed or both unringed, when adjoining the reflection 3 that interchanges them enlarges the group to

Then P is seen to be the $i\ldots j$-ambo orthoplex, where $i,\ldots,j$ are those nodes from 0,1,2 that were ringed in the original diagram.

In a similar way, we can adjoin an extra symmetry to see that the three polytopes

are also obtainable from the polyoctahedral group, as

The reducible reflection groups give rise to two kinds of prisms: those with one node disconnected from Coxeter diagram for a three-dimensional Archimedean or regular polyhedron, and the "proprism" (product prism) with Coxeter diagram

$$\underset{}{\odot}\!\!\overset{p}{\rule{1.2cm}{0.4pt}}\!\!\bullet \qquad \underset{}{\odot}\!\!\overset{q}{\rule{1.2cm}{0.4pt}}\!\!\bullet$$

a portion of which appears thus (with $p = 10$, $q = 15$):

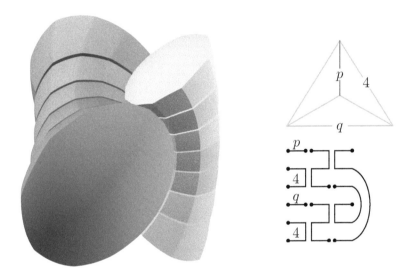

On the following pages, we illustrate the Archimedean polytopes arising by Wythoff's construction on the reducible four-dimensional reflection groups. The vertex figure and ring notation for the general polytope are followed by the generalized Schläfli symbol as in Chapter 21. Where the valence of a vertex or the number of sides of a face depends on the diagram, we have indicated this as (p), (q), or (r).

We then illustrate portions of the polytopes that the construction produces from the most interesting group $[5, 3, 3]$ in stereographic projection, accompanied by names in the ambo notation, and a list of the polyhedra at each vertex, as in Chapter 21; we also name the analogous polyhedra in the groups $[4, 3, 3]$, $[3, 3, 3]$, and $[3, 4, 3]$.

Regular Polytopes

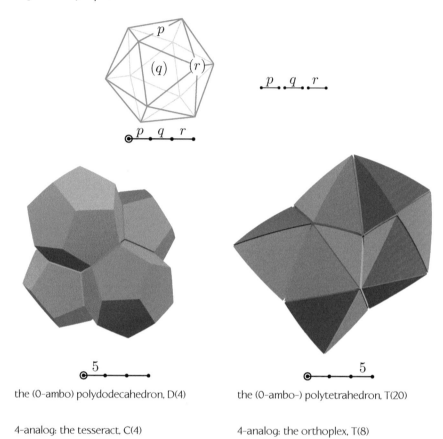

the (0-ambo) polydodecahedron, D(4)

4-analog: the tesseract, C(4)

the (0-ambo-) polytetrahedron, T(20)

4-analog: the orthoplex, T(8)

common 3-analog: the simplex, T(4)

further analog: the polyoctahedron, O(6)

1-ambo Polytopes

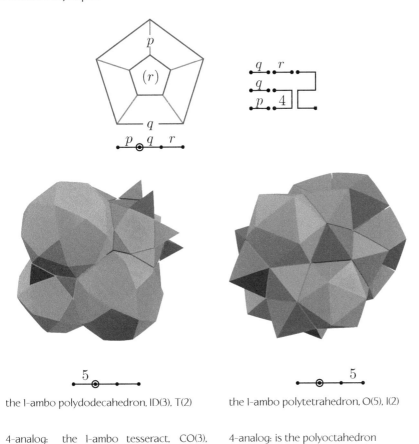

the 1-ambo polydodecahedron, ID(3), T(2)   the 1-ambo polytetrahedron, O(5), I(2)

4-analog: the   1-ambo   tesseract,   CO(3),   4-analog: is the polyoctahedron
T(2)

common 3-analog: the 1-ambo simplex, O(3), T(2)
further analog: the 1-ambo polyoctahedron, CO(3), C(2)

Truncated or 0l-ambo Polytopes

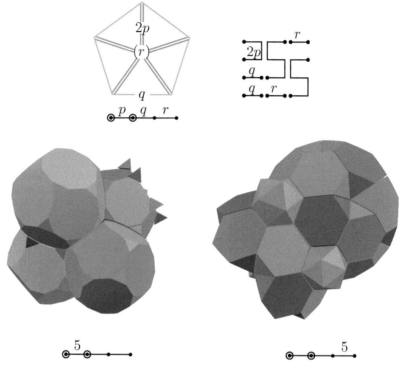

the 0l-ambo (or truncated) polydodecahedron, tD(3), T(l)

the 0l-ambo (or truncated) polytetrahedron,
tT(5), I(l)

4-analog: the 0l-ambo (or truncated)
tesseract, tC(3), T(l)

4-analog: the 0l-ambo (or truncated)
orthoplex, tT(4), O(l)

common 3-analog: the 0l-ambo (or truncated) simplex, tT(3), T(l)
further analog: the 0l-ambo (or truncated) polyoctahedron, tO(3), C(l)

Bitruncated or 12-ambo Polytopes

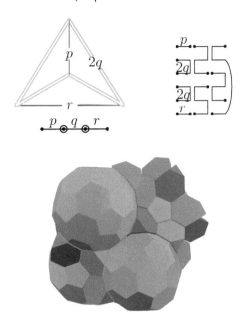

the 12-ambo (or bitruncated) polydodecahedron or polytetrahedron, tI(2), tT(2)

4-analog: the 12-ambo (or bitruncated) tesseract or orthoplex, tO(2),tT(2)

3-analog: the 12-ambo (or bitruncated) simplex, tT(4)

further analog: the 12-ambo (or bitruncated) polyoctahedron, tC(4)

02-ambo Polytopes

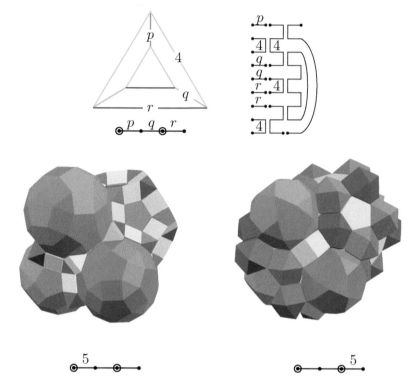

the 02-ambo polydodecahedron, RID(2), P3(2), O(1)

the 02-ambo polytetrahedon, CO(2), P5(2), ID(1)

4-analog: the 02-ambo tesseract, RCO(2), P3(2), O(1)

4-analog: is the 1-ambo polyoctahedron

common 3-analog: the 02-ambo simplex, CO(2), P3(2), O(1)

further analog: the 02-ambo polyoctahedron, RCO(2), P3(2), CO(1)

03-ambo Polytopes

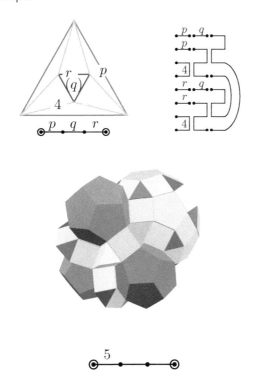

the 03-ambo polydodecahedron or polytetrahedron, D(l),P5(3), P3(3), T(l)

4-analog: the 03-ambo tesseract or orthoplex, C(1+3), P3(3), T(l)

3-analog: the 03-ambo simplex, T(2), P3(6)

further analog: the 03-ambo polyoctahedron, O(2), P3(8)

012-ambo Polytopes

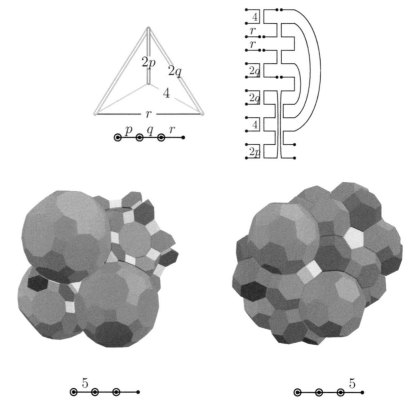

the 012-ambo polydodecahedron, tID(2), P3(I), tT(I)

the 012-ambo polytetrahedron, tO(2), P5(I), tI(I)

4-analog: the 012-ambo tesseract, tCO(2), P3(I), tT(I)

4-analog: is the 01-ambo (or truncated) polyoctahedron

common 3-analog: the 012-ambo simplex, tO(2), P3(I), tT(I)

further analog: the 012-ambo polyoctahedron, tCO(2), P3(I), tC(I)

O13-ambo Polytopes

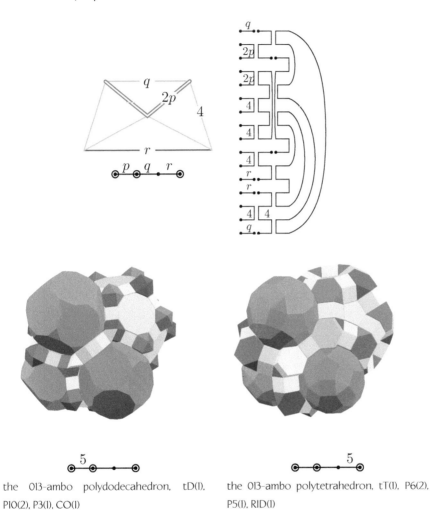

the O13-ambo polydodecahedron, tD(I), P10(2), P3(I), CO(I)

the O13-ambo polytetrahedron, tT(I), P6(2), P5(I), RID(I)

4-analog: the O13-ambo tesseract, tC(I), P8(2), P3(I), CO(I)

4-analog: the O13-ambo orthoplex, tT(I), P6(2), C(I), RCO(I)

common 3-analog: the O13-ambo simplex, tT(I), P6(2), P3(I), CO(I)

further analog: the O13-ambo polyoctahedron, tO(I), P6(2), P3(I), RCO(I)

Fully Expanded or 0123-ambo Polytopes

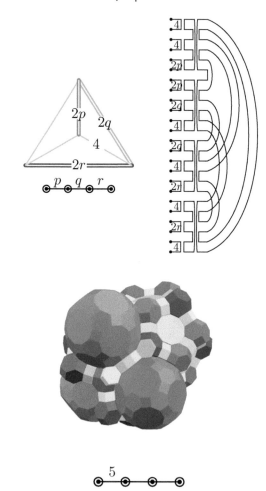

the 0123-ambo polydodecahedron or polytetrahedron, tID(1), P10(1),P6(1), tO(1)

4-analog: the 0123-ambo tesseract or orthoplex, tCO(1), P8(1), P6(1), tO(1)

3-analog: the 0123-ambo simplex, tO(2), P6(2)

further analog: the 0123-ambo polyoctahedron, tCO(2), P6(2)

The polytopes not obtainable from Wythoff's construction are, first, the prisms on the three-dimensional antiprisms and snub polyhedra, for which we give only the vertex figures

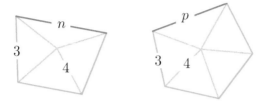

The vertex figures for the prisms on the three-dimensional antiprisms (left, with $n \geq 3$) and snub polyhedra (right, with $p = 4, 5$).

and, second, two very interesting polytopes, the "semi-snub polyoctahedron," discovered in 1900 by Thorold Gosset, and the "grand antiprism" found by Conway and Guy in 1965.

Gosset's Semi-snub Polyoctahedron

Coxeter called this polytope the "snub 24-cell," but since the operation here is really "semi-snubbing," our formal name is "semi-snub polyoctahedron." The cells of the 24-cell are octahedra, with a cubical vertex figure, a few of which are shown here, with a nice six-coloring.

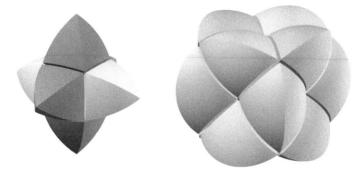

An icosahedron can be inscribed within each octahedron of the 24-cell

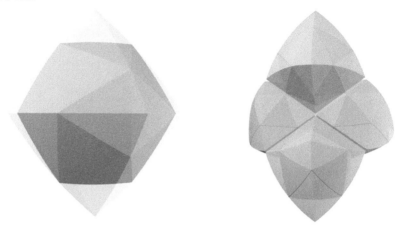

and clusters of five tetrahedra can replace each of the 24-cell's vertices, producing Gosset's semi-snub polytope. The 600-cell is produced in turn by replacing each of the twenty-four icosahedra with twenty tetrahedra.

The Grand Antiprism

This figure emerged only from the Conway–Guy enumeration and, like the Gosset polytope, can be described as the convex hull of some vertices of $\{3, 3, 5\}$. This time the omitted vertices form two equatorial decagons that lie in orthogonal planes. This replaces 300 of the original 600 tetrahedral cells by 20 pentagonal antiprisms, one for each of the omitted vertices.

There are two interlocked rings of pentagonal antiprisms in the grand antiprism, with three layers of tetrahedra between them. The tetrahedra of the middle layer, shown below on the right, have one edge on antiprisms of each ring.

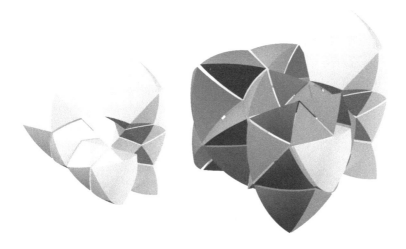

All three layers of tetrahedra are shown in the next figure; the vertex figure and Schläfli symbol appear on the right.

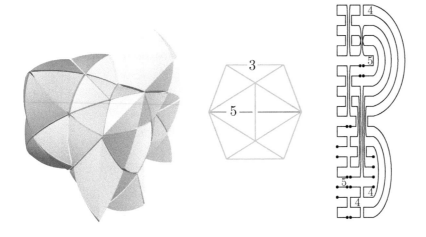

Regular Star-Polytopes

Regular star-polytopes exist only in two, three, and four dimensions. By rights the two- and three-dimensional ones should have appeared earlier in our book, but we have put them here as to incorporate them with the discussion of the four-dimensional ones.

Dimension 2

The regular star-polygons were discussed by Thomas Bradwardine in the fourteenth century. A typical one with Schläfli symbol $\{\frac{n}{d}\}$ has the same vertices as the regular n-gon, but its edges connect vertices that are d steps apart, where $1 < d < n/2$ and $(d, n) = 1$.

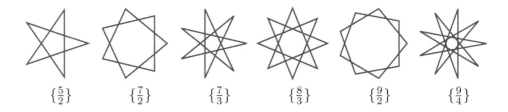

$\{\frac{5}{2}\}$ $\qquad$ $\{\frac{7}{2}\}$ $\qquad$ $\{\frac{7}{3}\}$ $\qquad$ $\{\frac{8}{3}\}$ $\qquad$ $\{\frac{9}{2}\}$ $\qquad$ $\{\frac{9}{4}\}$

The *density* of $\{\frac{n}{d}\}$ is d, meaning that its edges surround the center d times.

Dimension 3

In three-dimensional space, there are only four star-regular polytopes in three dimensions, namely the two discovered by Kepler early in the seventeeth century,

$\{\frac{5}{2}, 5\}$ the stellated dodecahedron,

$\{\frac{5}{2}, 3\}$ the great stellated dodecahedron,

and their duals, discovered by Poinsot in the nineteenth century,

$\{5, \frac{5}{2}\}$ the great dodecahedron,

$\{3, \frac{5}{2}\}$ the great icosahedron.

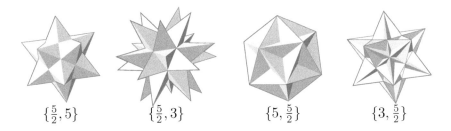

$\left\{\frac{5}{2},5\right\}$ $\left\{\frac{5}{2},3\right\}$ $\left\{5,\frac{5}{2}\right\}$ $\left\{3,\frac{5}{2}\right\}$

The list was proved complete shortly afterwards by Cauchy. The hexagon of Figure 26.1 illustrates all the relationships between all six pentagonal polyhedra. Polyhedra on the same horizontal line (joined by brown arrows) are mutually dual, while a blue arrow *stellates* a polyhedron, that is to say, takes it to another polyhedron whose edges are longer segments lying in the same lines.

Similarly, a green arrow *greatens* a polyhedron to another one whose faces are larger polygons of the same shape that lie in the same planes. The algebraic conjugation that changes the sign of $\sqrt{5}$ in all coordinates interchanges 5 with $\frac{5}{2}$ in the Schläfli symbols and takes each vertex of the hexagon to its opposite.

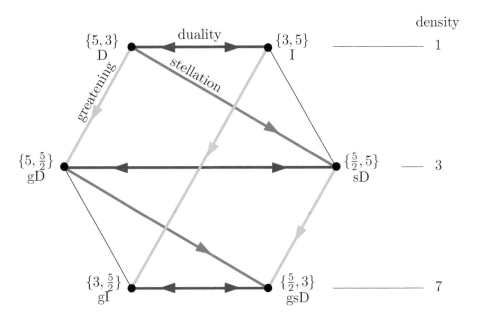

Figure 26.1. Relationships among the three-dimensional star-polytopes.

The four polyhedra of density 1 or 7 are topological spheres, but the two ($\{5, \frac{5}{2}\}$ and $\{\frac{5}{2}, 5\}$) of density 3 are surfaces of genus 4 since they are orientable of Euler characteristic $12 - 30 + 12 = -6$.

Dimension 4

Similarly, in four dimensions, the two convex pentagonal polytopes $\{5, 3, 3\}$ and $\{3, 3, 5\}$ are extended by the ten star-regular polytopes to a set of twelve that correspond to the vertices of a cuboctahedron.

The colored lines indicate various operations: brown for duality, blue for s = stellation, green for g = greatening, and a new operation red for a = aggrandisement, which replaces the three-dimensional cells by larger ones of the same shape lying in the same 3-spaces. Once again, the $\sqrt{5}$ conjugation replaces each vertex by its opposite.

In the symbols at the nodes are abbreviations of our suggested names:

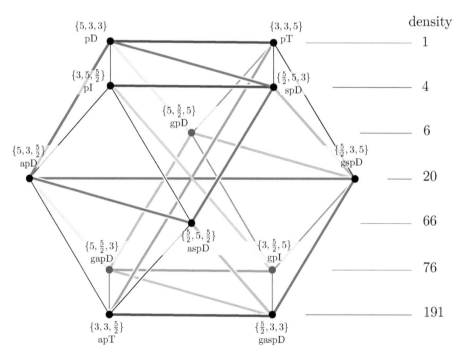

Figure 26.2. Relationships among the four-dimensional starry polytopes.

"g"   should be pronounced "great,"
"a"   should be pronounced "grand,"
"s"   should be pronounced "stellated,"
"p"   should be pronounced "poly-,"
"D"   should be pronounced "dodecahedron,"
"T"   should be pronounced "tetrahedron,"
"I"   should be pronounced "icosahedron."

Thus, $\{\frac{5}{2}, 3, 3\}$ is

"The great grand stellated polydodecahedron" (gaspD)!

Logically, the small letters g, a, s, and p can be written in any order; for instance, this is also the grand poly-(great stellated dodecahedron) since its cells are great stellated dodecahedra that are larger than those of the poly-(great stellated dodecahedron) but lie in the same 3-spaces.

All of these polytopes were discovered by Schläfli, but he rejected some of them on the grounds that they were not manifolds, as required by his definition. The ones that are manifolds are the six outermost ones in Figure 26.2. The others have 5's adjacent to $\frac{5}{2}$'s in the Schläfli symbol and so have points whose neighborhood is a cone on the genus-4 surface $\{5, \frac{5}{2}\}$ or $\{\frac{5}{2}, 5\}$. The only non-spherical 3-manifold that arises is the interesting one common to $\{5, 3, \frac{5}{2}\}$ and $\{\frac{5}{2}, 3, 5\}$.

Coordinates

The famous *Golden Number* τ is $(\sqrt{5}+1)/2$ with inverse $\sigma = (\sqrt{5} - 1)/2 = \tau - 1$, and the *Golden Ring* is $\mathbb{Z}[\tau]$. The *Icosians* are certain quaternions. We write $[a, b, c, d]$ for the Icosian $(a + bi + cj + dk)/2$, whose *norm* is $(a^2 + b^2 + c^2 + d^2)/4$. The definition of an Icosian is that its coordinates a, b, c, d must be in the Golden Ring and congruent modulo 2 *either* to each other *or* to some even permutation of $0, 1, \sigma, \tau$.

There are precisely 120 Icosians of norm 1, namely the even permutations of

$$A = [\pm 2, 0, 0, 0], \quad B = [\pm 1, \pm 1, \pm 1, \pm 1], \quad C = [0, \pm 1, \pm \sigma, \pm \tau],$$

which are the vertices of a polytetrahedron. (A alone yields an orthoplex, B a tesseract, A and B together a polyoctahedron.) In a similar way, the 600 Icosians of norm 2 are the vertices of a polydodecahedron. We write {norm} for the set of all Icosians of a given norm.

In this notation, the vertices and the centers of the edges, faces, and cells of all twelve pentagonal regular 4-polytopes are given below. Note that the set of vectors is always proportional to one of $\{1\}$, $\{2\}$, $\{\tau\sqrt{5}\}$, or $\{3\}$, which have respectively 120, 600, 720, or 1200 elements.

Schläfli Symbol	Vertices	Edge Centers	Face Centers	Cell Centers	Abbrev. Name
$\{5,3,3\}$	$\{2\}$	$\{3\sigma^2\}/2$	$\{\sigma\sqrt{5}\}/5$	$\{1\}/2$	pD
$\{3,3,5\}$	$\{1\}$	$\{\tau\sqrt{5}\}/2$	$\{3\tau^2\}/3$	$\{2\}/4$	pT
$\{3,5,\frac{5}{2}\}$	$\{1\}$	$\{\tau\sqrt{5}\}/2$	$\{3\tau^2\}/3$	$\{\tau^2\}/2$	pI
$\{\frac{5}{2},5,3\}$	$\{1\}$	$\{3\}/2$	$\{\tau\sqrt{5}\}/\sqrt{5}$	$\{\tau^2\}/2$	spD
$\{5,\frac{5}{2},5\}$	$\{1\}$	$\{\tau\sqrt{5}\}/2$	$\{\tau\sqrt{5}\}/\sqrt{5}$	$\{\tau^2\}/2$	gpD
$\{5,3,\frac{5}{2}\}$	$\{1\}$	$\{\tau\sqrt{5}\}/2$	$\{\tau\sqrt{5}\}/\sqrt{5}$	$\{1\}/2$	aspD
$\{\frac{5}{2},3,5\}$	$\{1\}$	$\{\sigma\sqrt{5}\}/2$	$\{\sigma\sqrt{5}\}/\sqrt{5}$	$\{1\}/2$	gpD
$\{\frac{5}{2},5,5\}$	$\{1\}$	$\{\sigma\sqrt{5}\}/2$	$\{\sigma\sqrt{5}\}/\sqrt{5}$	$\{\sigma^2\}/2$	aspD
$\{5,\frac{5}{2},3\}$	$\{1\}$	$\{3\}/2$	$\{\sigma\sqrt{5}\}/\sqrt{5}$	$\{\sigma^2\}/2$	gapD
$\{3,\frac{5}{2},5\}$	$\{1\}$	$\{\sigma\sqrt{5}\}/2$	$\{3\sigma^2\}/3$	$\{\sigma^2\}/2$	gpI
$\{3,3,\frac{5}{2}\}$	$\{1\}$	$\{\sigma\sqrt{5}\}/2$	$\{3\sigma^2\}/3$	$\{2\}/4$	apT
$\{\frac{5}{2},3,3\}$	$\{2\}$	$\{3\tau^2\}/2$	$\{\tau\sqrt{5}\}/5$	$\{1\}/2$	gaspD

Star-Archimedean Things

There are also starry analogs of the Archimedean polyhedra, which were tentatively classified in a beautiful 1954 paper by Coxeter, Miller, and Longuet-Higgins. Their list was proved complete by Skilling, with the exception of an interesting but slightly problematic polyhedron he discovered.

So far as we know, nobody has yet enumerated the analogs in four or higher dimensions.

Since the only finite reflection groups in five dimensions are those of the regular polytopes (see below), the five-dimensional polytopes given by Wythoff's construction all have ambo-names:

$ij\ldots k$-ambo simplex, or

$ij\ldots k$-ambo orthoplex $= IJ\ldots K$-ambo hypercube, where now

$i + I = j + J = \ldots = k + K = 4.$

A similar statement is true in nine or higher dimensions. However, in six, seven and eight dimensions, there are some fascinating new polytopes corresponding to the exceptional reflection groups E_6, E_7, and E_8 (see below).

Groups Generated by Reflections

This is the largest geometrically-defined class of groups that have been completely enumerated in all dimensions. According to Coxeter's famous theorems, the groups correspond to *diagrams* known as Coxeter graphs (or sometimes as Dynkin diagrams).

The diagrams for the irreducible Euclidean groups are given in Figure 26.3. The reducible groups are direct sums of the irreducible ones.

There is an interesting correspondence between these and some of the irreducible spherical groups, the so-called "crystallographic" ones. These are obtained by omitting the (solid) *extending nodes*, and their names are obtained by removing the tilde (so that $B_n = C_n$.) However, there are a few non-crystallographic irreducible spherical groups, whose diagrams we now give:

Unfortunately, the theory of reflection groups is too rich in both content and application for us to pursue it further here. However, we shall briefly mention some objects associated with the most interesting reflection groups.

Hemicubes

The n-dimensional *hemicube* is the convex hull of alternate vertices of the n-dimensional hypercube. It can also be obtained by Wythoff's

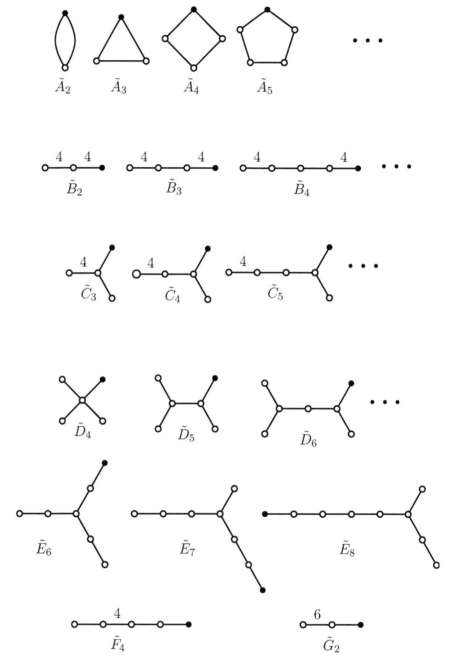

Figure 26.3. Coxeter graphs of the irreducible Euclidean groups.

construction, and here are the Coxeter-Wythoff symbols for the first few cases:

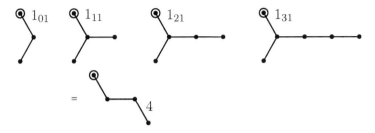

These symbols explain Coxeter's notation for the $(n+3)$-dimensional hemicube, 1_{n1}. The first few cases are rather special. We start with $n = 0$ because the "two-dimensional" hemicube is just a line segment. The Coxeter-Wythoff symbol shows that the three-dimensional hemicube is just a tetrahedron, while the four-dimensional one is an orthoplex, but with only half its symmetry, as is better revealed by the generalized Schläfli symbols:

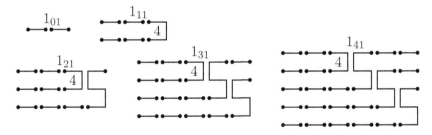

These show that in n dimensions the flag-rank is $n - 2$ and there are two types of cell, a simplex and the $(n - 1)$-dimensional hemicube.

The five-dimensional hemicube is special in that both its cells are regular polyhedra, the 4-simplex and 4-orthoplex, although the latter appears with only half symmetry. It is a member of a more interesting series of polytopes, as follows.

The Gosset Series

Gosset discovered a much more interesting, but finite, series of polytopes, all of which have flag-rank 3. The first few are shown in the following figure.

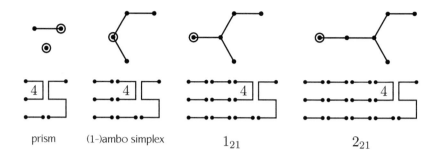

prism   (1-)ambo simplex 1_{21} 2_{21}

The disconnected Coxeter-Wythoff symbol in the three-dimensional case shows that this is a Cartesian product of a triangle and an interval, i.e., a triangular prism.

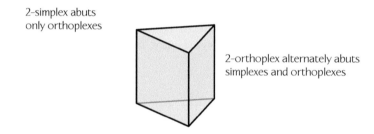

2-simplex abuts
only orthoplexes

2-orthoplex alternately abuts
simplexes and orthoplexes

The combinatorics of this case apply to all members of the Gosset series; in every case, their cells are simplexes and orthoplexes, the latter appearing with only half symmetry. Each simplex abuts only orthoplexes, while the orthoplex cells alternately abut simplexes and other orthoplexes.

The Coxeter-Wythoff symbol of the four-dimensional case shows that it is the four-dimensional (1-)ambo simplex, whose vertices are the edge midpoints of a 4-simplex. In this case the orthoplexes arise as ambotetrahedra (i.e., octahedra) while the simplex cells are the tetrahedra originating from the truncated vertices.

As we have already remarked, the five-dimensional case is the 5-hemicube 1_{21}. The six-dimensional case is the interesting Hesse polytope (2_{21} in Coxeter's notation), which has 72 simplex cells and 27 orthoplex cells opposite its 27 vertices.

There are only three more cases. The Hesse polytope 3_{21}, the Gosset polytope 4_{21}, and the Gosset tessellation 5_{21}.

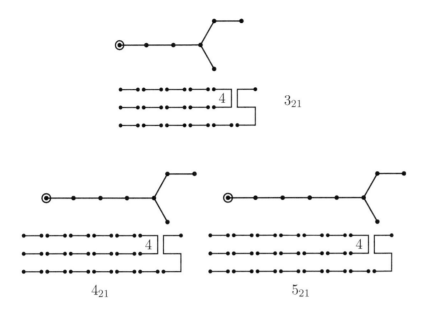

3_{21}

4_{21} 5_{21}

For $n \leq 4$, it is natural to regard n_{21} as a tessellation on a spherical surface of dimension $n+3$. However, 5_{21} is a tessellation of *Euclidean* 8-space by simplexes and orthoplexes, and so its symmetry group is infinite. Its vertices form the celebrated E_8 root lattice.

The Gosset tessellation is vaguely analogous to the tessellation of 3-space that we called tetroctahedrille (or tetroctille) except that the orthoplexes in the Gosset tessellation abut both simplexes and orthoplexes.

Although Gosset did not discover it, there is in fact one further term to the series: the tessellation 6_{21} of hyperbolic 9-space by "ideal" simplexes and orthoplexes having all their vertices at infinity!

The Symmetries of Still Higher Things

There is a wonderful packing of spheres in eight dimensions, almost certainly the densest, in which the spheres are centered at the points of the E_8 lattice. In 1965 John Leech discovered a similar packing of spheres in 24 dimensions, again almost certainly the densest centered at the points of what is now called the Leech lattice, Λ_{24}. (Cohn and Kumar have proved that *no* packings in dimensions 8 and 24 can be more than microscopically denser than those given by E_8 and Λ_{24}, which are the densest *lattice* packings.)

One of us made his mathematical name by proving that the number of symmetries of the Λ_{24} packing that fix any given sphere is

$$8,315,553,613,086,720,000.$$

(Of course, the entire symmetry group is infinite.)

This is not all. In 1973 Berndt Fischer discovered, and in 1980 Bob Griess constructed, the so-called Monster group, which arises as the symmetries of a 196883-dimensional polytope! Its order is

$$808,017,424,794,512,875,886,459,904,961,710,757,005,754,368,000,000,000$$

Where Are We?

We may seem to have come a long way, but in fact we have barely scratched the surface of the mathematics of symmetry. A universe awaits—Go forth!

– A –

Other Notations for the Plane and Spherical Groups

The columns of Table A.1 correspond to different notation systems, subject to the remarks noted below. The column titles are abbreviations:

OS	our orbifold signature
I	International notation
C&M	Coxeter and Moser
S	Speiser
N	Niggli
P	Pólya
G	Guggenheim
F	Fejes Tóth
C	Cadwell

Our orbifold signature is the one presented in this book. The International notation is the most used of the older notations. The C&M notation is the notation used in Coxeter and Moser's *Generators and Relations for Discrete Groups* [11], which should be consulted for the individual references.

The notations p3m1 and p1m3 were inadvertently interchanged by Niggli, whose notation is otherwise taken from Spieser with the addition of the Roman numerals in parentheses. This error is repeated in editions of Coxeter and Moser before 1980, by which time Doris Schattschneider [24] and H. Martyn Cundy [12] had independently discovered the interchange, and in many other places. We thank Schattschneider for this information.

OS	I (C&M)	S (N)	P G	F C
*632	p6m	$C_{6v}^{(I)}$	D_6	W_6^1
632	p6	$C_6^{(I)}$	C_6	W_6
*442	p4m	$C_4^{(I)}$	D_4^*	W_4^1
4*2	p4g	C_{4v}^{II}	D_4°	W_4^2
442	p4	$C_4^{(I)}$	C_4	W_6
*333	p3m1	C_{3v}^{II}	D_3^*	W_3^1
3*3	p31m	C_{3v}^{I}	D_3°	W_3^2
333	p3	$C_3^{(I)}$	C_3	W_3
*2222	pmm	C_{2v}^{I}	D_2kkkk	W_2^2
2*22	cmm	C_{2v}^{IV}	D_2kgkg	W_2^1
22*	pmg	C_{2v}^{III}	D_2kkgg	W_2^3
22×	pgg	C_{2v}^{II}	D_2gggg	W_2^4
2222	p2	$C_2^{(I)}$	C_2	W_2
**	pm	C_s^{I}	D_1kk	W_1^2
*×	cm	C_s^{III}	D_1kg	W_1^1
××	pg	C_2^{II}	D_1gg	W_1^3
○	p1	$C_1^{(I)}$	C_1	W_1

Table A.1. The Euclidean plane groups.

OS	C	S	W	P&M	I
*532	$[3,5]$	I_h	$\bar{P}$	I	$\bar{5}3m$
532	$[3,5]^+$	I	P	I	532
*432	$[3,4]$	O_h	$\bar{W}$	O_i	$m3m$
432	$[3,4]^+$	O	W	O	432
*332	$[3,3]$	T_d	WT	T_O	$\bar{4}3m$
3*2	$[3^+,4]$	T_h	$\bar{T}$	T_i	$m3$
332	$[3,3]^+$	T	T	T	23
*22N	$[2,N]$	D_{Nh}	$\bar{D}_N$	$D_N i$	N/mmm or $2\bar{N}\,m2$
2*N	$[2^+,2N]$	D_{Nd}	$D_{2N}D_N$	$D_N\,D_{2N}$	$2\bar{N}\,2m$ or $\bar{N}\,m$
22N	$[2,N]^+$	D_N	D_N	D_N	N2
*NN	$[N]$	C_{Nv}	$D_N\,C_N$	$C_N\,D_N$	Nm
N*	$[2,N^+]$	C_{Nh}	$\bar{C}_N$	$C_N i$	N/m or $2\bar{N}$
N×	$[2^+,2N^+]$	S_{2N}	$C_{2N}C_N$	$C_N\,C_{2N}$	$2\bar{N}$ or $\bar{N}$
NN	$[N]^+$	C_N	C_N	C_N	N

Table A.2. The spherical groups.

The abbreviations for Table A.2 are as follows:

OS	our orbifold signature
C	Coxeter
S	Schoenflies
W	Weyl
P&M	Pólya and Meyer
I	International notation

Table A.2 compares our signature with older notations for the spherical groups; it is adapted from Coxeter and Moser's *Generators and Relations for Discrete Groups*, which should again be consulted for the references. The reader should be warned that the fonts have been uniformized for simplicity and that for N = 1 or 2 there are various special notations and equivalences that we have ignored, since they become obvious from the signature when digits 1 are omitted. Some pairs of lines contain notations in braces, which as they stand are for even values of N, but should be interchanged when N is odd.

Bibliography

[1] Hans Ulrich Besche, Bettina Eick, and Eamonn O'Brien. "The Groups of Order at Most 2000." *Electron. Res. Announc. Amer. Math. Soc,* 7:1–4, 2001.

[2] H. Brown, R. Bülow, J. Neubüser, H. Wondratsschek, and H. Zassenhaus. *Crystallographic Groups of Four-Dimensional Space.* John Wiley and Sons, New York, 1978.

[3] Peter Buser, John Conway, Peter Doyle, and Klaus-Dieter Semmler. "Some Planar Isospectral Domains." *Int. Math. Res. Notes,* 9:391–400, 1994.

[4] John H. Conway. "The Orbifold Notation for Surface Groups." In M. W. Liebeck and J. Saxl, editors, *Groups, Combinatorics and Geometry: Proceedings of the L.M.S. Durham Symposium, July 5-5, Durham, U.K., 1990,* London Mathematical Society Lecture Note Series 165, pages 438–447, Cambridge, 1992. Cambridge University Press.

[5] John H. Conway, Heiko Dietrich, and E.A. O'Brien. "Counting Groups: Gnus, Moas and Other Exotic Creatures." To appear.

[6] John H. Conway, Olaf Delgado Friedrichs, Daniel H. Huson, and William P. Thurston. "On Three-Dimensional Space Groups." *Contributions to Algebra and Geometry,* 42(2):475–507, 2001.

[7] John H. Conway and Michael J. T. Guy. "Four-Dimensional Archimedean Polytopes." In *Proceedings of the Colloquium on Convexity, Copenhagen 1965,* pages 38–39. Kobenhavns University Mathematics Institute, Copenhagen, 1967.

[8] John H. Conway and Juan Pablo Rossetti. "Describing the Platycosms." Eprint at arXiv:math/0311476, 2003.

[9] John H. Conway assisted by Francis Y.C. Fung. *The Sensual (Quadratic) Form.* The Carus Mathematical Monographs Number 26. The Mathematical Association of America, Washington, D.C., 1997.

[10] H.S.M. Coxeter. "Regular Skew Polyhedra in Three and Four Dimensions and Their Topological Analogues." *Twelve Geometric Essays*, pages 75–105. Southern Illinois University, Carbondale, 1968.

[11] H.S.M. Coxeter and W. Moser. *Generators and Relations for Discrete Groups*. Ergebnisse der Matematik und Ihrer Grenzgebrete 14. Springer, Berline, 1957. (Later editions published in 1965, 1972, and 1980.)

[12] H. Martyn Cundy. "p3m1 or p31m?" *Mathematical Gazette*, 63:192, 1979.

[13] Ludwig Danzer. Personal communication, 1996.

[14] Carolyn Gordon, David L. Webb, and Scott Wolpert. "One Cannot Hear the Shape of a Drum." *Bulletin of the American Mathematical Society*, 27:134–138, 1992.

[15] Thorold Gosset. "On the Regular and Semi-regular Figures in Space of n Dimensions." *Messenger of Mathematics*, 29:43–48, 1900.

[16] Edouard Goursat. "Sur les substitutions orthogonales et les divisions régulières de l'espace." *Annales Scientifiques de L'École Normale Supérieure*, 6:9–102, 1889.

[17] Branko Grünbaum and G. C. Shephard. "Spherical Tilings with Transitivity Properties." *The Geometic Vein*, pages 65–98. Springer-Verlag, New York, 1982.

[18] Branko Grünbaum and Geoffrey Shephard. *Tilings and Patterns*. W.H. Freeman & Company, New York, 1986.

[19] Graham Higman. "Enumerating p-groups I: Inequalities." *Proc. London Math Soc.*, 10(3):24–30, 1960.

[20] Mark Kac. "Can One Hear the Shape of a Drum? Part II." *American Math. Monthly*, 73(4):1–23, 1966.

[21] John Morgan and Gang Tian. *Ricci Flow and the Poincaré Conjecture*. Clay Mathematics Monographcs 3. AMS Press, New York, 2007.

[22] W. Nowacki. "Über die Anzahl verschiedener Raumgruppen." *Schweiz. Mineral. Petrogr. Mitt.*, 34:130–168, 1954.

[23] László Pyber. "Asymptotic Results for Permutation Groups." *Amer. Math. Soc. DIMACS Series*, 11:197–219, 1993.

[24] Doris Schattschneider. "The Plane Symmetry Groups, Their Recognition, and Notation." *American Mathematical Monthly*, 85:439–450, 1978.

[25] Charles C. Sims. "Enumerating p-groups." *Proc. London Math Soc.*, 15(3):151–166, 1965.

[26] Derek Smith and John H. Conway. *Quaternions and Octonians.* A K Peters, Ltd., Natick, MA, 2003.

[27] T. Sunada. "Riemannian Coverings and Isospectral Manifolds." *Ann. of Math.*, 121:169–186, 1985.

[28] W. Threlfall and H. Seifert. "Topologische Untersuchung der Diskontinuitätsberieche endlicher Bewegungsgruppen des dreidimensionalen sphärischen Raumes." *Mathematische Annalen*, 104:1–70, 1931.

[29] P. Du Val. *Homographies, Quaternions, and Rotations.* Oxford University Press, Oxford, UK, 1964.

[30] J. R. Weeks and G. K. Francis. "Conway's Zip Proof." *American Mathematical Monthly*, 106:393–399, 1999.

[31] J.R. Weeks. *The Shape of Space.* Pure and Applied Mathematics. Marcel Dekker, New York, 2002.

Index

T - #0753 - 101024 - C444 - 235/191/20 [22] - CB - 9781568812205 - Gloss Lamination